COVID-19 PANDEMIC 2019-2020

An International & Personal Guide to a Pandemic

Dedicated to all the First Responders

Dedicated to: _____

_____
NIK NIKAM, MD, MHA

NIK NIKAM, MD, MHA

COVID-19 PANDEMIC 2019-2020

Copyright: S. G. (NIK) NIKAM, MD, MHA
All rights reserved ©2020
Editor: L. Pay, Ph.D.

Cover Graphics: NIK NIKAM, MD, MHA
Cataloging in Publication Date Oct. 2020
NIKAM, NIK
COVID-19 PANDEMIC 2019-2020
First eBook: 2020

Version 1.0

Publication Date: NOV 15, 2020
ISBN: 9780976527572
EAN 13: 978-0-9765275-7-2
Library of Congress Control Number: 2019901233
Language: English

NIK NIKAM, MD MHA
NNN Media
3130 GRANTS BLVD #17034
SUGAR LAND, TX 77496
<u>drniknikam@gmail.com</u>

Printed in the United States of America

COVID-19 PANDEMIC 2019-2020

An International & Personal Guide to a Pandemic

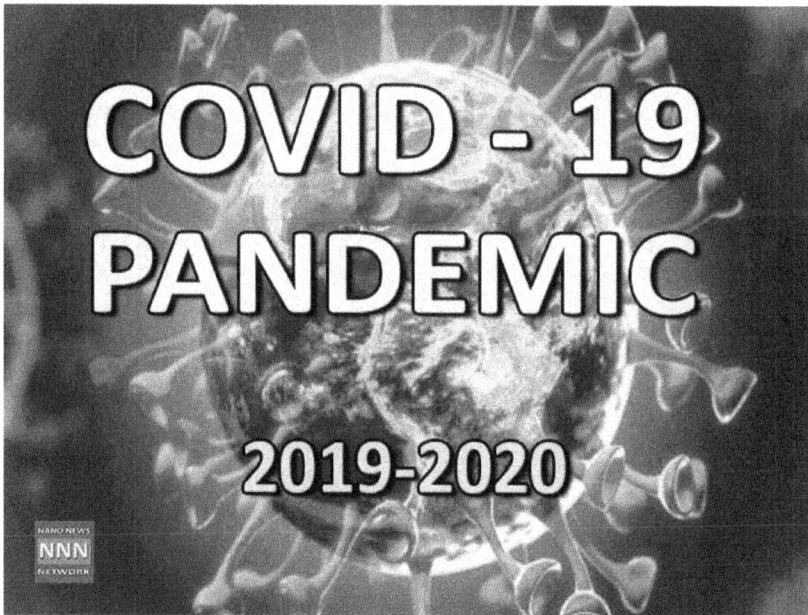

NIK NIKAM, MD, MHA

Publications by Nik Nikam, MD, MHA

Stressless Mind & Priceless Body (1995). Paperback.

Heart-Healthy Lifestyle (2010). Paperback and Kindle Editions.

RAMAYAN – An English Screenplay (2012). Paperback and Kindle Editions.

Cruise Crisis – An English Screenplay (2013). Kindle Edition.

Stressfree Lifestyle (2016). Paperback and Kindle Editions.

Clinical EKG Interpretation (2018) Paperback.

Cardiology Lecture Series:

https://www.youtube.com/channel/UCL1x8qabFk8zXn6XOHq8w1w

YOUTUBE: NIK NIKAM

Please visit our YouTube channel, "NIK NIKAM," and subscribe to it.

COVID-19 PANDEMIC 2019-2020

Author: NIK NIKAM, MD, MHA

Dr. Nik Nikam has practiced cardiology in Houston and Sugar Land, Texas, for over 40 years.

He has had vast exposure to the medical field at the world-famous Texas Medical Center for over three decades. His public speaking skills and interactions with diverse populations have contributed to a lasting reservoir of prolific scientific and general knowledge, experience, and wisdom.

Dr. Nikam has assembled immense knowledge on the COVID-19 pandemic from around the world from the time it appeared in Wuhan, China. He has written extensively on social media about various aspects of the COVID-19 pandemic, including the incubation period, the Farr curve, the spread of infection, clinical presentations, treatment options, mortality, hospitalization trends, antibodies, vaccines, lockdown pros and cons, the paradigm shifts from office to home office, and the possibility of a second wave. He emphasizes the lessons learned at every step, and how these could be applied to slow the viral spread in the preliminary stages of epidemics to save lives and protect jobs and the economy.

This book will be an indispensable guide in the event of a new epidemic or pandemic and serves as a reference guide for future world leaders with advice on generating a stockpile of PPE, developing new testing devices, maximizing healthcare resources, flattening the curve, implicating self-mitigation guidelines, containing the spread, and staying ahead of the Farr curve to save lives. Dr. Nikam explains these essential steps and many other approaches to managing a pandemic in simple language to provide a valuable resource for individuals, small business owners, schoolteachers, healthcare advisors, state public health officials, and policymakers.

Dr. Nikam is a professional speaker, a Distinguished Toastmaster (DTM), an author, writer, auctioneer, and talk-show host. His YouTube channel, NIK NIKAM NETWORK (NNN), has over 25,000 subscribers and over 1400 videos on important medical, cultural, and educational topics.

Dr. Nikam enjoys delivering engaging and persuasive presentations sprinkled with humor. For speaking engagements, contact Dr. Nik Nikam at:

NIK NIKAM, MD, MHA
3130 GRANTS BLVD #17034
SUGAR LAND, TX 77496
281-745-4161
drniknikam@gmail.com

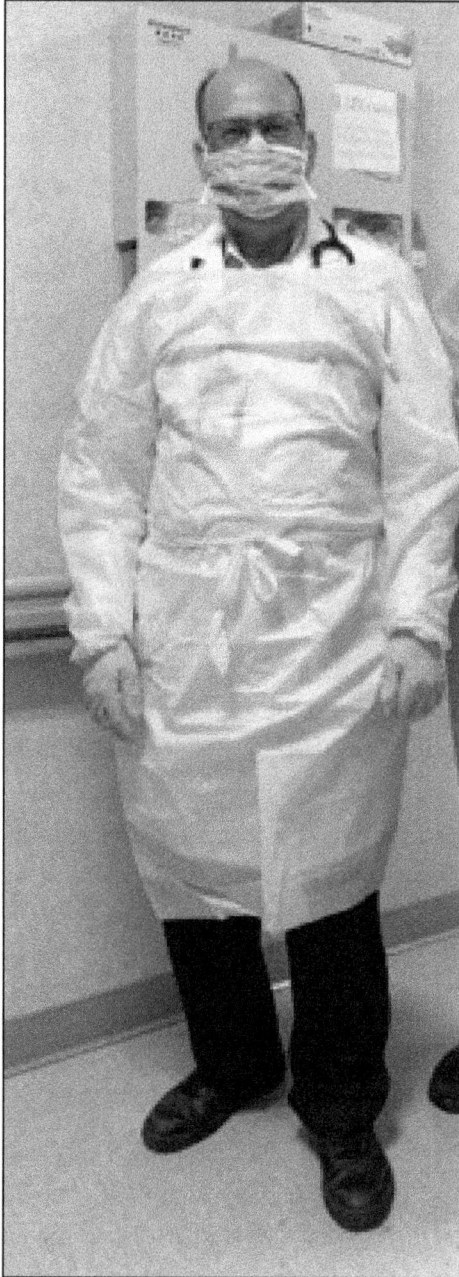

1 *Table of Contents*

Sermon before you start this journey!

John, a farmer, takes his preacher on a tractor to show off his fields. As John drives around, the preacher is impressed by acres and acres of beautifully cultivated land.

The preacher looks at John and says, "John, you and the good Lord have made it here together."

John replies, "Yes, brother, you should have seen it when your good Lord had it by himself!"

***

Hopefully, this book will help you to cultivate approaches for dealing with the next epidemic or a pandemic.

ACKNOWLEDGMENTS:

Most of the graphs are from www.worldometers.com

Images are from google images.

Thanks to so many who inspired me to write a series of social media articles in educating ourselves in a rapidly evolving and fleeing course of this COVID-19 pandemic with an uncertain future.

My personal thanks to L. Pay, Ph.D., for her exceptional editing skills, great insight on the topic, and timely response.

1 INTRODUCTION

"Houston, we have a problem!"

Those were the words of Jack Swigert to the NASA control center in Houston from Apollo 13 in 1993—words that reverberate in our memories and come alive whenever we see calamities in and around our lives. We have somehow learned to live through such challenges, frequently overcoming deadly hurricanes and other disasters.

Little did we know, in 2020, the arrival of COVID-19 would spark a global pandemic with unprecedented magnitude.

Houston has been my hometown for the past four decades. I have fond memories of Houston, with its bugs, humidity (no wonder it earned the nickname 'Humbug City,' or simply 'H town'), hurricanes, and, of course, the NASA space center.

I moved from Little Rock, Arkansas to Temple, Texas, on March 7, 2020, to start my new Cardiology job on March 29. I had three weeks of solitary confinement in a tiny apartment in a town no bigger than my former neighborhood in Fort Bend county.

My passion for public speaking and interest in medical journalism inspired me to create the NNN media channel on YouTube, which has over 25,000 subscribers and more than 1,400 videos covering our H Town lives.

Technical analysis in the stock market always fascinated me; particularly, diamonds (DIA), triple Q (QQQ), and spiders (SPY). I followed the rise and fall of the curves that followed a set pattern. Certain technical pointers point to what the next move could be based on historical data. Similarly, pandemics go through cycles that follow a certain set pattern. The curves generated by the rise and fall of the number of new cases or deaths are referred to as the Farr curve, which is similar to the bell curve we see in medical studies and population distributions etc.

When I saw Wuhan's COVID-19 cases on a time graph, it gripped my attention and curiosity. I felt like I had seen something like that before. Then, I looked at the Italian curve, and later, the New York curve, a couple of weeks behind Italy's. I started working on possible projections based on my technical knowledge of the stock market and the Bell curve we had studied in scientific research classes. Comparing the graphs from the other two countries, I imagined what might happen in the United States in the coming days and weeks.

I intensely studied:

✓ How many days did it take for new case numbers to reach the top of the Farr curve from the time the first case was announced?
✓ How long did case numbers stay at the top of the "plateau," as they call it?
✓ How many days did it take for case numbers to reach the bottom of the curve?

This was the beginning. The images of Wuhan flashed in my mind, sending chills down my spine as I remembered thousands of people in Wuhan walking with masks. The entire city of eleven million people in a lockdown state. All travels to and from Wuhan halted. Next, images of deathly sick Italians unable to access care due to a shortage of ICU beds.

We fail to appreciate the gravity of a situation until it hits home.

COVID-19 PANDEMIC 2019-2020

We all love New York as it is the symbol of America to people all over the world. When cases began to rise rapidly in New York City, it became clear: COVID-19 had invaded the United States of America.

I am not an infectious disease specialist, or a pulmonary & critical care specialist, or an epidemiologist. I resemble a restless child, curious about everything in life: why, how, what, when, where? I am innately drawn to unanswered questions. You can imagine the result of locking up a restless mind in a tiny apartment: thus, this book, "COVID-19 Pandemic: An International and Personal Guide to a Pandemic," was born.

In my quest for answers, I researched day and night as a medical journalist only to find out that we knew little about the COVID-19 pandemic, much less how to treat the viral illness and contain it from spreading. While making the last additions to this chapter, I realized I had covered over 35 topics related to the COVID-19 pandemic while it was still carving out its first wave in many parts of the world.

Hopefully, many chapters in this book will serve as a guide for dealing with future pandemics including how people should anticipate an oncoming pandemic, what they should prepare before the landfall, which tests are needed to detect new viruses, and why decisive actions must be taken from day one to prevent viral spread, save lives, and preserve the economy

I hope this book not only serves as a historical rendition of the COVID-19 pandemic as I saw it, but also helps international leaders, policymakers, healthcare professionals, and public health experts learn how to prevent viral spread and save lives, restore the economy, find a cure, and live another day to tell their favorite stories!

***

Severe acute respiratory syndrome coronavirus (SARS-CoV-2) is a coronavirus that causes the respiratory illness known as Coronavirus Disease 19 (COVID-19). SARS-CoV-2 was first identified in December 2019 in Wuhan, Hubei, China, and rapidly evolved into a pandemic involving over 200 countries.

COVID-19 PANDEMIC 2019-2020

The first confirmed case was traced back to Nov 17, 2019, in Hubei. As of October 6, 2020, the total number of COVID-19 cases worldwide had climbed above 35.8 million, with over 1.05 million deaths and 26.9 million recoveries. From China, it spread to the middle east, with the greatest number of cases occurring in Iran. (Figure 1-1)

Then, it spread to Europe in early March, affecting Germany, France, Italy, Sweden, Austria, and many other countries.

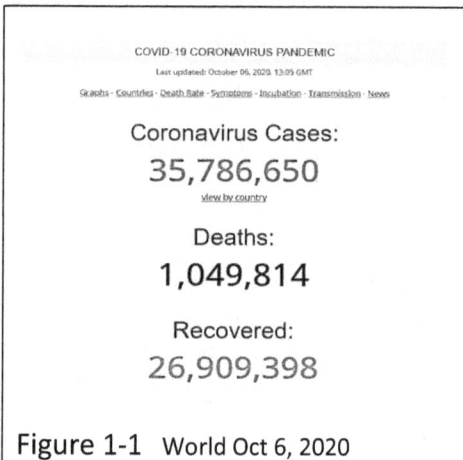

COVID-19 CORONAVIRUS PANDEMIC
Last updated: October 06, 2020, 13:05 GMT
Graphs · Countries · Death Rate · Symptoms · Incubation · Transmission · News

Coronavirus Cases:
35,786,650
view by country

Deaths:
1,049,814

Recovered:
26,909,398

Figure 1-1 World Oct 6, 2020

It quickly spread to the US and Canada in the middle of March, where New York City became the epicenter of the COVID-19 pandemic.

Then, COVID-19 spread to India in the middle of March and to South America in the later weeks of April. The COVID-19 pandemic forced most countries into a lockdown state for 4-6 weeks, including India, where more than 1.3 billion people were plunged into lockdown.

This book is a historical reflection of how I saw the world react and respond to the COVID-19 pandemic while I was under house arrest (my fun expression for lockdown!) for weeks, sitting in front of my computer for hours and days at a time, losing sight of date, time, and place. This may not be a very good statement to make as a physician— as medical professionals, the first thing we establish with any new patient is whether that person is oriented to date, time, place, and who the President of the country is... At least I knew that Donald J. Trump was the President of the US at the time of this writing as I watched most of his White House Coronavirus Briefings while—you guessed it—sitting alone in front of my computer monitor.

I dedicate this book to all the healthcare workers who knowingly and willingly took risks to save the lives of COVID-19 patients, knowing that they could lose their own in the process. It reminds me of the first batch of firefighters who rushed to the Chernobyl nuclear accident site in 1986, though perhaps not to that extent.

I salute all the world leaders for taking the necessary steps to do their best (yes, some of you might take an issue with this) to protect their citizens, save lives, prevent the spread of viruses, and provide financial assistance of unparalleled magnitude to their people.

There was no textbook on how to deal with the COVID-19 pandemic. Everyone based their plans on knowledge from past pandemics and advice from world-leading authorities on infectious diseases and pandemics, economic experts, and passionate policymakers.

I applaud leaders who showed extortionary courage, vision, empathy, and sympathy, bringing the corporate world and services together to serve the millions in times of crisis.

This book is not intended to be a factually accurate rendition of the facts, for there is no such thing as a fact. Beauty is in the eye of the beholder! The so-called facts depend on which media people watch and which political spectrum they belong to, injecting their own pre-conceived or ill-conceived notions of what the world looks like through their myopic eyes, and what it should look like in their make-believe world to suit their philosophies. This book is not about that.

It is not even a purely a scientific diatribe on COVID-19 etiology, incubation, presentation, symptoms, signs, laboratory tests, differential diagnoses, treatment, prognosis, and postmortem analysis. Instead, this is a living document of how I, as a medical journalist, with a touch of my medical background as a Cardiologist for over 40 years, personally saw the COVID-19 pandemic unfold before our eyes in living colors and horrors.

Also, you might not see the eloquence of the King's English in this document as the rapidly gushing thoughts in my swiftly fleeting mind were translated into text with extremely limited typing skills. I should say I tried voice-to-text, but maybe it's my old-fashioned approach of typing and seeing the thoughts transform into logical (sometimes illogical) blocks one sentence at a time. It's like laying one brick at a time while imagining the finished Taj Mahal!

Throughout this book, I have attempted to document what we learned from each experience as we accumulated a vast amount of information, and how the wisdom learned from this devastating human experiment in the 21st century might be applied to deal with future similar challenges.

The chapter on Wuhan, the city in China where the COVID-19 pandemic began, looks at how COVID-19 originated and what effects it had on that population of eleven million people. The lessons learned from Wuhan laid the foundation for what the rest of the world could expect as the virus spread outside of China like a wild Californian forest fire.

COVID-19 is unlikely to kill us, but the associated depression, isolation, lack of physical activity, job stress, and economy might very well cause crippling long-term health, economic, and social issues. The economic and psychological effects of pandemics have been well documented. People lose jobs, spouses, children, money, houses, and gain weight, depression, anxiety, addiction, and potentially, a wasted life. It is vital that we learn how to act to avoid this devastating effect.

FARR CURVE OR BELL CURVE

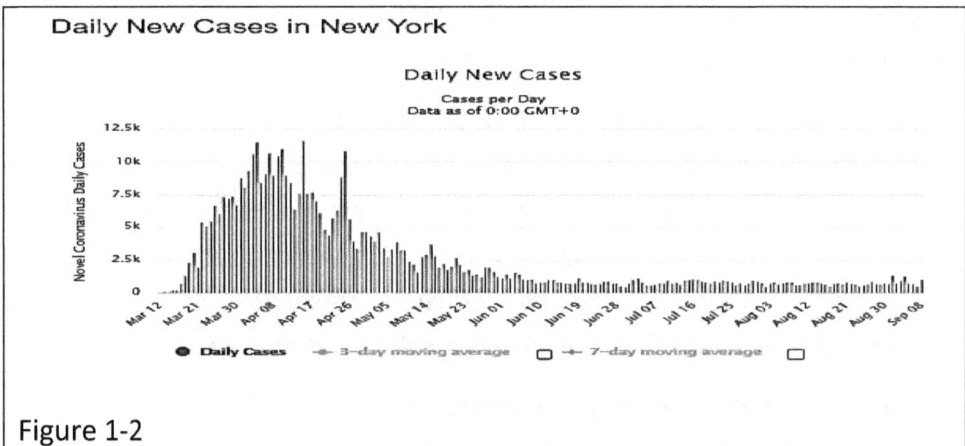

Figure 1-2

When you look at an epidemic or pandemic curve, you will notice a similarity to the 'Bell curve.' There is an incubation period from isolated case reports to a sudden surge in the number of new cases, hospitalizations, ICU use, or deaths. Next, the curve reaches a peak. It can take days or weeks to reach the peak, which is influenced by the local conditions, the organism's virulent nature, and mitigation rules put in place. The plateau that follows may last for days or weeks. Finally, a steady decline follows the plateau, occurring much more slowly than the initial brisk rise in numbers. Eventually, the numbers reach their low point. This is referred to as the Farr Curve.

In cases of widespread pandemics, this low point may never reach zero. While a pandemic may end with a small number of cases reported from time to time, it is also possible that a second and third wave similar to the rippling effects of the first wave may occur. The emergence of these second and third spikes has been attributed to people ignoring self-mitigation rules, or the emergence of a more infectious form of the virus that spreads to a much larger population.

How the virus manifests in different parts of the world reflects local conditions, individuals' comorbid conditions, population density, age, and the population's previous exposure to similar viruses. How various governments respond to a pandemic, with strict lockdowns and self-mitigation rules to no lockdowns or masks at all, also influences the Farr Curve.

The Farr curve upstroke of COVID-19 was of a shorter duration than the downward slope in almost all countries. The upstroke duration and height varied vastly from country to country. It lasted for less than two weeks in Wuhan, but over six months in India. The country's size also influenced the upstroke: New Zealand, with a population of five million people, had a much smaller upstroke than India, with its population of 1.3 billion. Hence, the daily count of new cases varied across countries from a few cases to 80,000 per day.

The downslope of the curve generally lasted two to three times longer than it took to reach the peak. In some instances, the counts never reached zero. In many parts of the world, a second surge arose (Texas, USA), while other countries were still climbing the first wave (India). People attributed second surges to too-early removal of lockdown restrictions, others defying the self-mitigation rules, and people going to the extent of intentionally getting the infection to get over the restrictions.

It was also clear from the various curves that most lockdowns were introduced at a time when the sudden surge in cases was in full swing. If lockdowns and self-mitigation had been introduced earlier in the epidemic in more countries, as was done in New Zealand, the number of cases and deaths could have been vastly reduced. This is an important lesson to take home for future pandemics.

THE SECOND AND THIRD WAVES

The 1918-1919 Spanish 'Flu had a second wave that was five times more devastating than the first one. It also lasted much longer. Then, a third wave followed before the pandemic eventually disappeared. Between 1918 and 1919, the Spanish 'Flu killed more than 50 million people worldwide.

Now, we have better means to deal with a viral pandemic; however, the assault cannot be escaped until the development of an effective vaccine, and without "flattening the curve" with lockdowns and mitigation procedures, human and economic damage that has the potential to devastate many countries is likely. Like a Tsunami or a Hurricane, a pandemic can wreak havoc on civilization. The first-line defense strategy in the initial stage of a pandemic is to flatten the Farr curve and bring down the initial upswing in the infection numbers, spreading cases over a longer period to prevent an overloading of the healthcare systems that could lead to millions of deaths. (Figure 1-2)

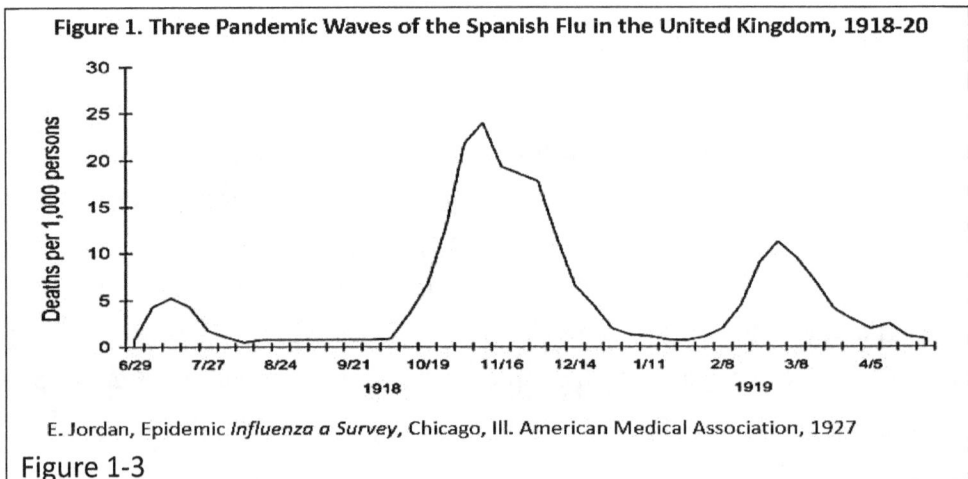

Figure 1. Three Pandemic Waves of the Spanish Flu in the United Kingdom, 1918-20

E. Jordan, Epidemic *Influenza a Survey*, Chicago, Ill. American Medical Association, 1927

Figure 1-3

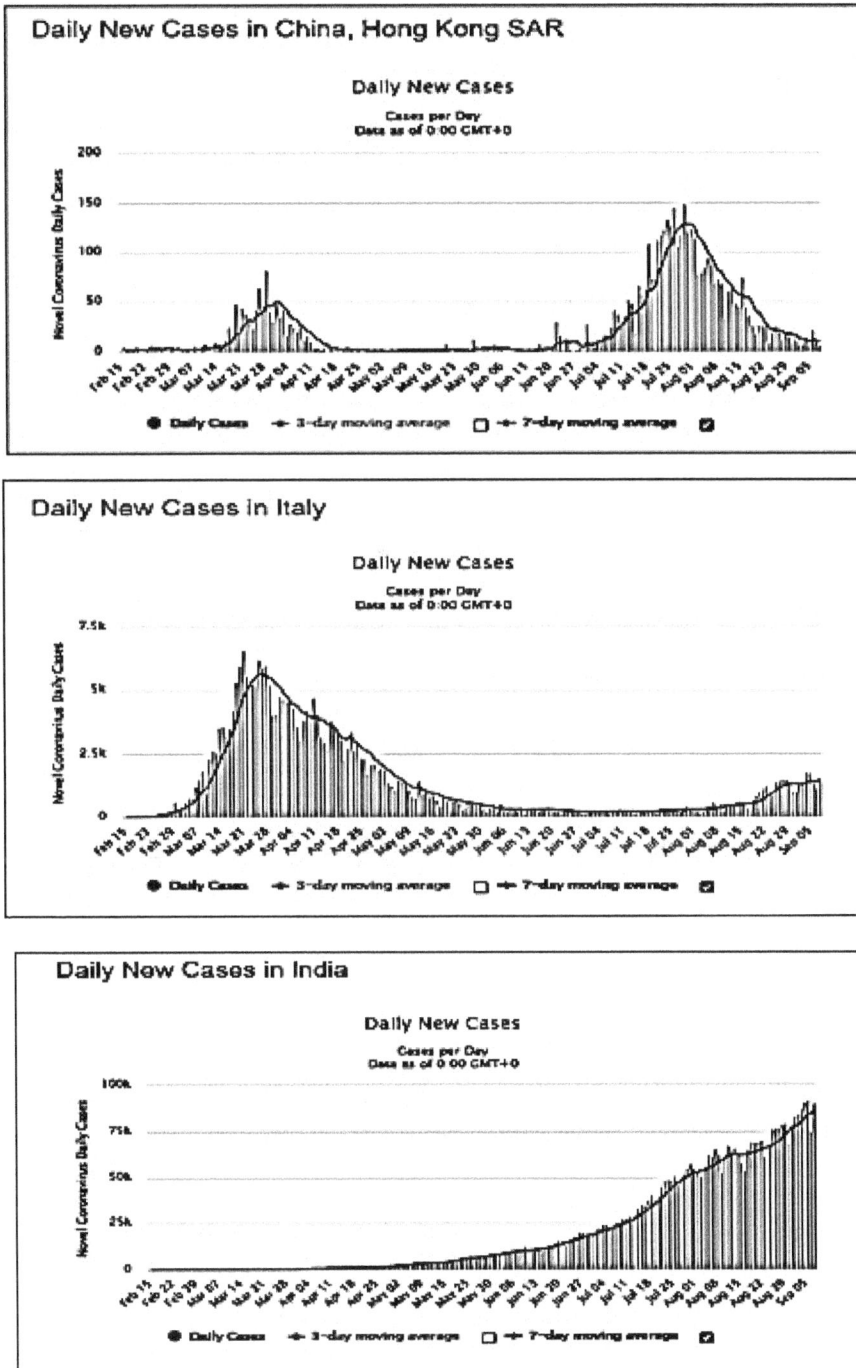

Figure 1-4 Ref: worldometers.com

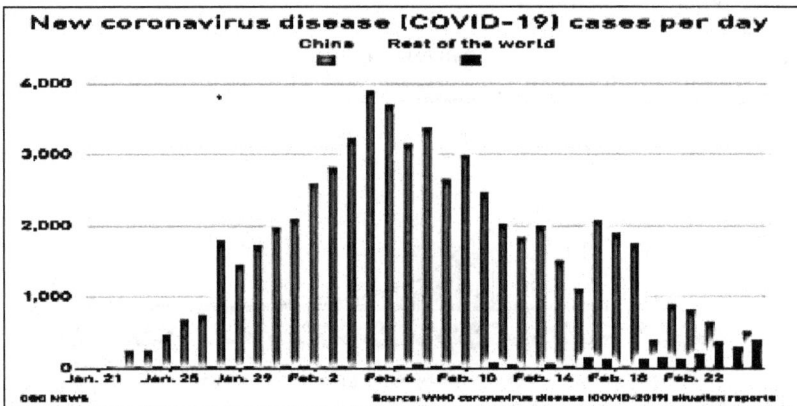

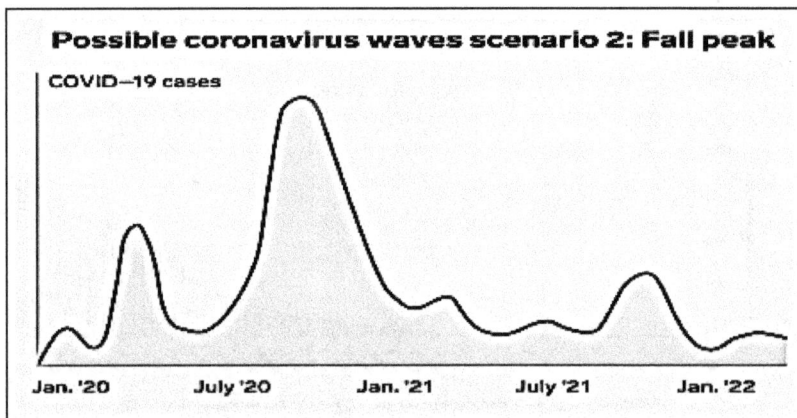

Figure 1-5 Ref: worldometers.com

2 NUMBERS OR TRENDS?

The COVID-19 pandemic, first reported in the news in January 2020, spread from its starting point in Wuhan, China, to over 200 countries, wreaking havoc worldwide with no sign of relief. How could world leaders respond to this relentless assault on human lives, depleting the healthcare resources around the world, and sinking the world into an economic calamity? What could they rely upon for guidance—**Numbers or Trends?**

NUMBERS

If you went by the numbers, the US had the highest number of cases and performed the greatest number of tests (87 million tests as of September 2020). More testing identifies more cases; thus, more active cases will be reported in a country able to perform a lot of tests than in a country in which testing is limited. How does that help a country learn about the direction and impact of the virus on their population if they are not able to test enough people to perform accurate comparisons with countries with widespread testing?

The US also had the most mortality. On September 7, they had 193,500 deaths per 6.46 million positive cases. That put the mortality from COVID-19 in the US at 0.29% of active cases.

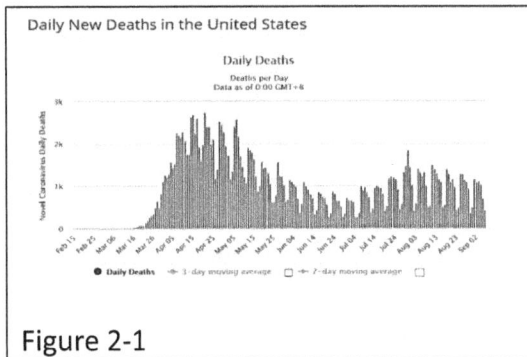

Daily New Deaths in the United States

Daily Deaths

Deaths per Day
Data as of 0:00 GMT+8

Figure 2-1

The mortality numbers were highly subjective as many deaths outside of hospitals were not accounted for. The way mortalities were recorded and reported was influenced by human error and manipulation. There were instances where death rates had to be modified when records were challenged, and where data were suppressed to look better on the world scene or inflated to get more federal government aid. It is important to note that leaders and countries might manipulate pandemic data for various political reasons; thus, numbers may not be the most reliable way to track a pandemic if the source of the numbers is not impartial.

INDEXING

The next approach was to index the death rates per million (M) so states or countries with different populations could be more accurately compared. However, that did not take into account the fact that COVID-19 was not evenly distributed throughout the country. New York City experienced 100 times more cases than some of the other states.

THE TABLE

It appears from the table below, the US had a mortality of 583/1M, while Brazil had 595/1M, Peru had 903/1M, and India had 52/1M. Those numbers do not reflect the extent of COVID-19 spread or the direction. The US graph showed that the country already had a second spike while India was still in its first. Ironically, the media could take those numbers and spin them in any manner to suit their audience, to push their missions, and their ideologies. So, the media interpretations were as confusing as the tables and numbers (Figure 2-2)

	Country, Other	Total Cases	Total Deaths	Tot Cases/ 1M pop	Deaths/ 1M pop	Total Tests	Tests/ 1M pop	Population
	World	27,283,718	887,305	3,500	113.8			
1	USA	6,460,250	193,250	19,496	583	87,474,919	263,986	331,362,224
2	India	4,202,562	71,687	3,040	52	48,831,145	35,321	1,382,493,246
3	Brazil	4,137,606	126,686	19,440	595	14,408,116	67,695	212,838,474
4	Russia	1,025,505	17,820	7,027	122	38,400,000	263,111	145,946,206
5	Peru	689,977	29,838	20,873	903	3,379,580	102,240	33,055,232
6	Colombia	666,521	21,412	13,074	420	2,936,517	57,599	50,982,171

(Tabs above table: All | Europe | North America | Asia | South America | Africa | Oceania)

Figure 2-2

PERCENTAGE GAME

Another approach was to compare the death rates to the country's population as a percentage of the world's population. One media source repeatedly tried to push the narrative that the US had 5% of the world's population and 25% of the deaths from the COVID-19 infections. That was erroneous data that was being propagated to misguide the public about the gravity of the situation—the entire population of the US was not infected with COVID-19, and many countries had no COVID-19 cases.

The hospitalization numbers reflected only part of the disastrous story. Many people infected with COVID-19 did not even reach the hospitals. Then, patients were recorded as COVID-19 cases despite not receiving a COVID-19 test. Many intentional and unintentional variables skewed those numbers. Some hospitals even inflated COVID-19 numbers to get more federal aid.

Even the world healthcare leaders and experts downplayed the seriousness of the pandemic, gave conflicting reports on the modes of transmission, provided limited and confusing guidance at the beginning of the pandemic, and promoted unrealistic projections.

If most of these numbers were manipulated and unreliable, what can we depend on to understand the severity of a disease, its trend, and its direction?

Weather forecasters look at trends, directions, trajectories, and make future projections. Trend analysis is essential to anticipate what to expect and to predict what could change the direction of a pandemic, flatten the curve, and reduce the demand on the health care system.

THE TREND IS YOUR FRIEND

The "trend is your friend" idiom, very widely applied in the stock market, was perhaps the best gauge on making sense of where the pandemic was on the Farr curve in any country, when to expect its peak, how to track the decline, and how to prepare for an onslaught of deathly sick people flocking the hospitals.

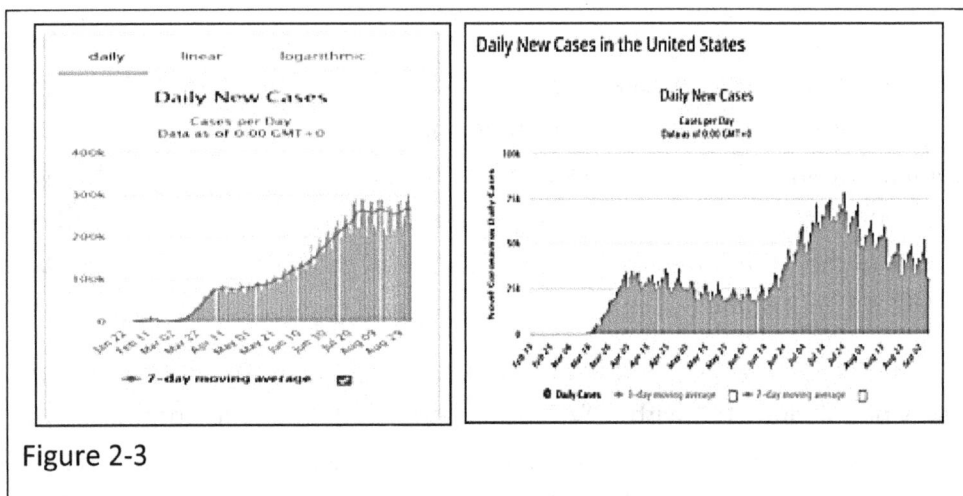

Figure 2-3

On the timeline, the number of new cases in the world looked different from that of the US. While the world had reached a peak at over 300,000 new cases, there were still no signs of a significant decline in the number of new cases. On the other hand, in the US, the number had already peaked and declined twice; however, the numbers had not yet reached the bottom of the curve.

RUSSIA, BRAZIL, AND INDIA

Russia, Brazil, and India marched right behind the US in numbers but sketched a different story of the COVID-19 pandemic. Even though these countries saw their first cases in the middle of March, as of September, their trends looked different.

RUSSIA reached a peak in the middle of May 2020. Then, it showed a steady decline. Yet, as of September 7, they were still recording over 5000 new cases per day. That number is high compared to the number of new cases in Hong Kong, but exceptionally low compared to new case numbers in India, which were ten times higher. Hence, the trend is your friend. (Figure 2-4)

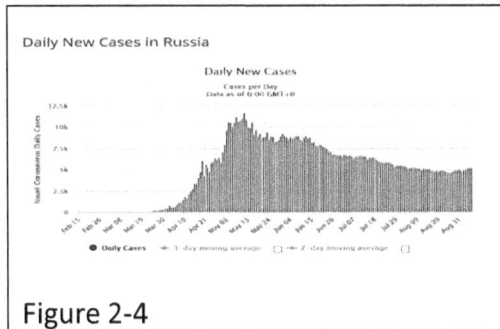

Figure 2-4

BRAZIL experienced its first case toward the end of March. At its peak in July, it had close to 70,000 new cases per day. Cases then steadily declined, and, as of September 7, their new cases were still close to 30,000 per day. Brazil's mortality rate was 595/1M, higher than that of the US, which stood at 583/1M. This trend suggests that numbers could either continue to decline or rise again in a second spike, which, based on past pandemic trends, might be more explosive than the first. (Figure 2-5)

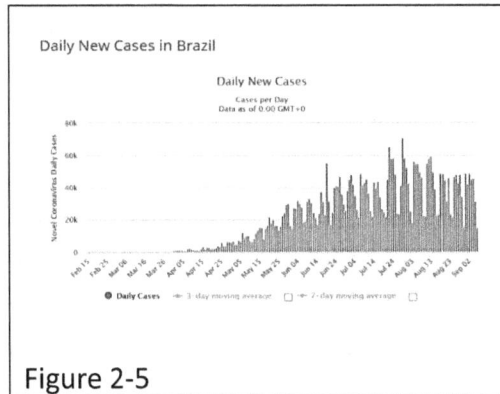

Figure 2-5

INDIA. In September, India became the epicenter of the COVID-19 pandemic. India paralleled the US in the number of tests performed (48.8 million v 87 million in the US) and the number of new cases, which stood at 4.2 million. On September 7, India had the second-highest number of cases in the world. Those numbers, based on its population size of 1.3 billion, may appear small. But don't let numbers deceive you! (Figure 2-6)

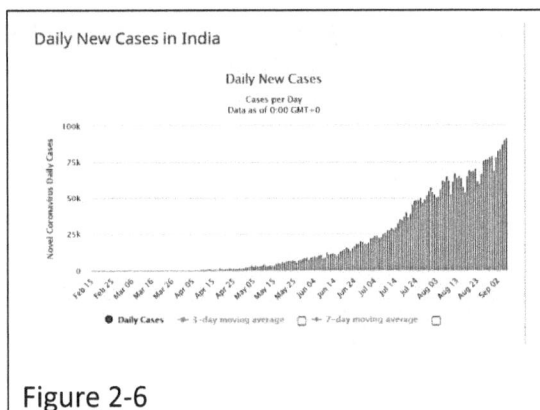

Figure 2-6

The **trend** told a different story. The graph indicates that India might not have reached its peak and, based on the Farr curve theory, had yet to plateau or begin to decline. India had more than 90,000 new cases per day at the beginning of September, and if it took two to three times longer for other countries to decline to a reasonable number from their peaks, it was reasonable to speculate that it could take several months to see those low numbers of new cases in India.

IF THE TREND IS A FRIEND, IS VACCINE THE ONLY RAY OF HOPE?

I am an optimistic person. If the challenges were grave and devastating, the frontier pioneers are ahead of the curve in developing a vaccine, which could be our hope for mass protection. Looking at the promising results from Phase 1 and 2 trials, and with more than half a dozen Phase 3 trials underway, there are hopes we could be within striking distance, perhaps months, from having a commercially available vaccine.

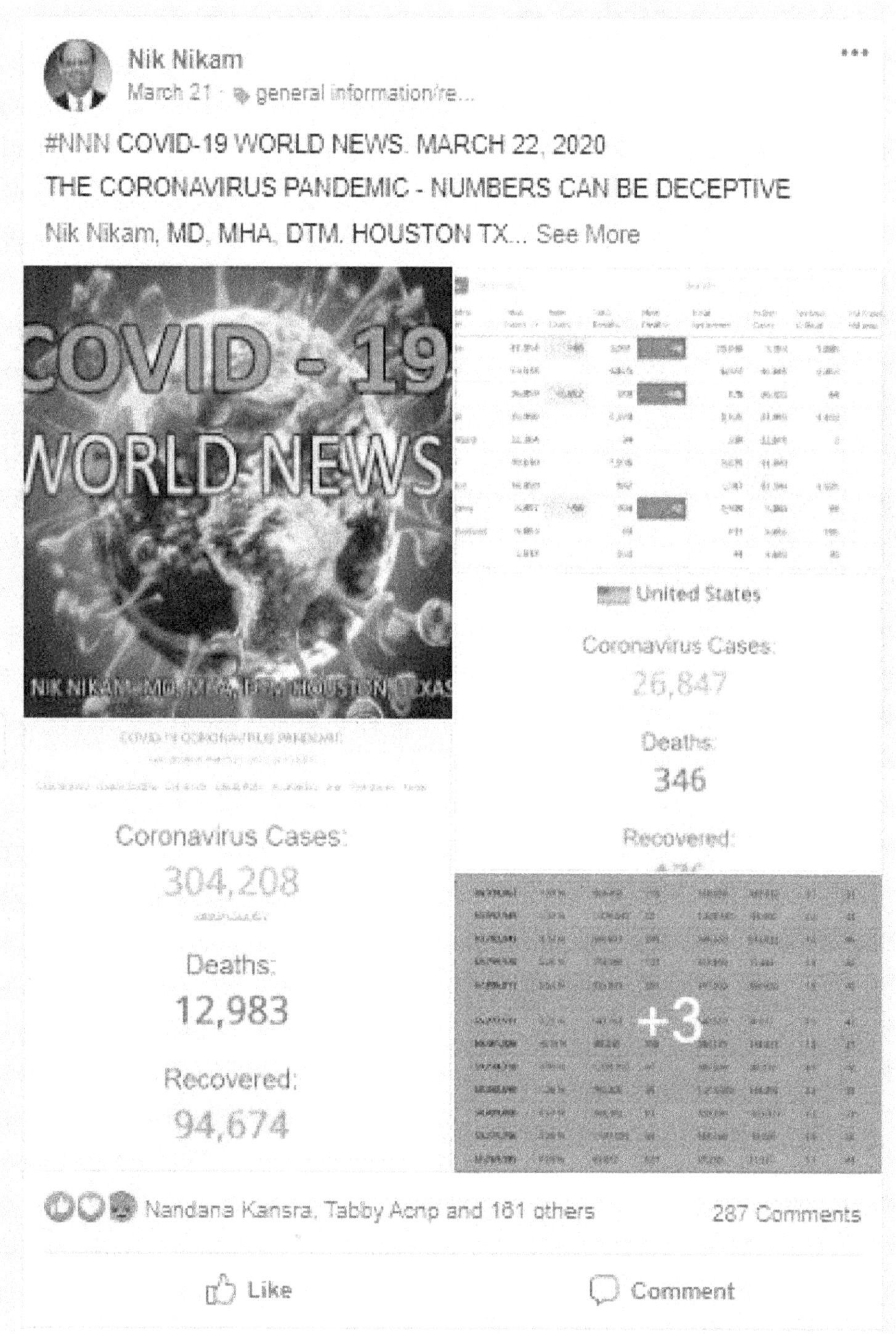

Nik Nikam
March 21 · general information/re...

#NNN COVID-19 WORLD NEWS. MARCH 22, 2020

THE CORONAVIRUS PANDEMIC - NUMBERS CAN BE DECEPTIVE

Nik Nikam, MD, MHA, DTM. HOUSTON TX... See More

Coronavirus Cases:

304,208

Deaths:

12,983

Recovered:

94,674

United States

Coronavirus Cases:

26,847

Deaths:

346

Recovered:

+3

OOO Nandana Kansra, Tabby Acnp and 181 others   287 Comments

👍 Like   💬 Comment

Nik Nikam
March 21 · general information/re...

#NNN COVID-19 WORLD NEWS. MARCH 22, 2020

THE CORONAVIRUS PANDEMIC - NUMBERS CAN BE DECEPTIVE

Nik Nikam, MD, MHA, DTM. HOUSTON TX... See More

COVID-19 WORLD NEWS

NIK NIKAM, MD, MHA, DTM, HOUSTON, TEXAS

Coronavirus Cases:
304,208

Deaths:
12,983

Recovered:
94,674

United States
Coronavirus Cases:
26,847

Deaths:
346

Recovered:

+3

Nandana Kansra, Tabby Acnp and 161 others   287 Comments

Like   Comment

3   PAST PANDEMICS

LESSONS FROM PAST PANDEMICS TO FUTURE PANDEMICS

Spanish 'Flu: The first wave started in July 1918. It was interesting that it began in the middle of the summer, as this is unusual for an Influenza virus. The first wave lasted 30 days, as did the initial COVID-19 cycle. Then, there was a gap of two months.

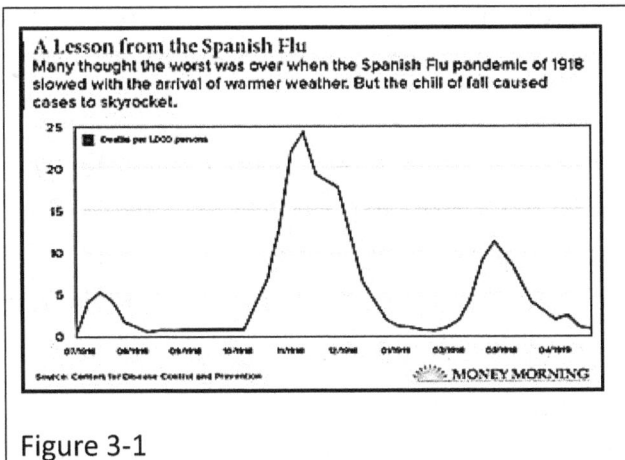

A Lesson from the Spanish Flu
Many thought the worst was over when the Spanish Flu pandemic of 1918 slowed with the arrival of warmer weather. But the chill of fall caused cases to skyrocket.

Source: Centers for Disease Control and Prevention   MONEY MORNING

Figure 3-1

The second wave came like a Tsunami, and death rates went up 500%. That cycle lasted three months. The third wave, though smaller than the second, was twice as devastating as the first and lasted two months. In the end, the total death toll worldwide was over 50 million. Yet, 75% of the world population survived the Spanish 'Flu. (Figure 3-1)

The Justinian plague between 451 and 452 AD killed 30-50 million people. The Bubonic plague between 1347 and 1351 AD lasted four years and killed more than 200 million people. Smallpox in 1520 killed 56 million people.

The H1N1 paralleled the 'flu seasons in Thailand between 2008 and 2009, lasting 18 months. (Figure 3-2)

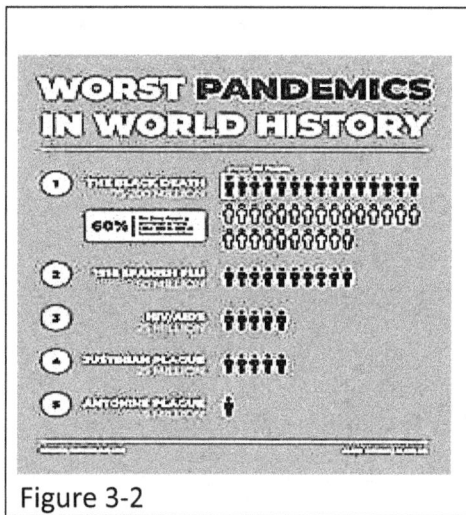

Figure 3-2

Based on these observations, what options were available in any new epidemic or pandemic?

We knew there would be no cure for a novel virus at the beginning. One option was to diagnose the active cases, isolate them, and quarantine them.

The second crucial step was to contain the spread of the new virus from infecting other people. Masks were highly effective in preventing viral spread from person to person. Often, we saw the implementation of masks or mandates requiring people to wear masks come too late, long after widespread viral infection. Even after recommending and mandating the masks, their implementation was not always successful in the US due to individual and community pushback against mask-wearing.

Countries such as South Korea, Hong Kong, Taiwan, New Zealand, Australia, and Japan, implemented mask mandates very early in their infection timeline and effectively educated their people on the importance of masks in preventing viral spread. They had remarkable success in minimizing viral spread, even without sustained lockdowns.

Mask use very early during a pandemic can significantly contain the virus, reduce the number of new cases, and decrease the number of hospitalizations. In many instances, countries with mask mandates were able to keep businesses open, even during peak epidemic season.

During the 1918 Spanish 'Flu, people in St. Louis practiced social distancing and flattened the curve significantly compared with that of Philadelphia, which did not implement such measures. It might not have reduced the overall mortality, but by reducing the demand on healthcare personnel and resources, St. Louis was able to avoid the chaos caused by an overwhelmed healthcare system, while Philadelphia was not.

Did self-quarantine mean kicking the ball down the timeline, and how long they could keep doing it?

As we knew from past epidemics, a second and third wave of COVID-19 was possible, and it was thought that this might arrive around the regular 'flu season, further complicating the situation.

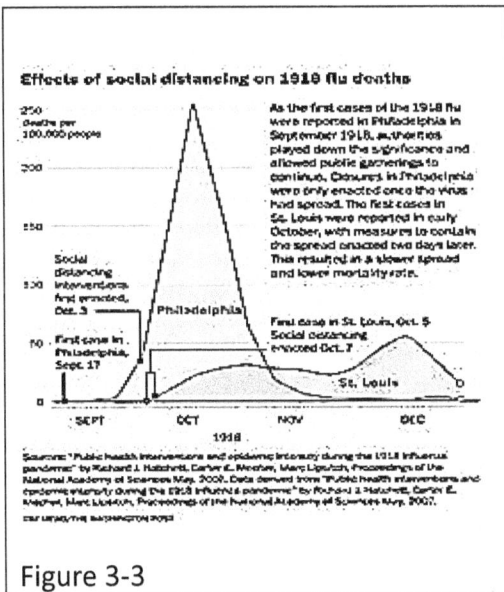

Figure 3-3

Without an extensive understanding of the extent of the immunity in the community to the COVID-19 virus, it is challenging to try new interventions that might result in additional side effects.

First, let us look at potential outcomes in different scenarios. (Figure 3-3)

One scenario was to continue the self-mitigation steps such as wearing a mask, social distancing, and frequent handwashing until the virus ran its course. At least, with these techniques, you might open businesses, start schools and colleges, and restore the economy.

An alternate scenario was that the virus would fade or lose its potency during the summer months and would not resurface again in the fall.

Alternately, a new vaccine might provide immunity to the world's population. However, by the time a vaccine became available, the COVID-19 pandemic might have already delivered a devastating blow in terms of human suffering, death, and economic calamity.

The worst scenario could be one where COVID-19 becomes as deadly as the 1918 Spanish 'Flu, resulting in millions of deaths. The entire Spanish 'Flu season lasted from July 1918 to May, 1919. That was ten months. That was not a timeframe in which anyone could develop a vaccine.

A daring option would be to go for herd immunity at the onset. It was a risky approach, which was partly adopted by Sweden. They recommended self-mitigation steps, but people were not required to follow them. They still had their businesses running. Sweden had extremely high mortality rates, though not the highest in Europe, and did not even come close to the herd immunity they were aiming for. When they tested for antibodies, only 10% to 14% of their population had developed antibodies.

When New York, the most affected state in the US, did random antibody testing in various parts of the state, they found 21% immunity in New York City. Let us assume for a moment that this number was accurate enough for our argument.

That means when you reopen businesses, the number of active COVID-19 cases might go up. But there was no way to predict how fast it would increase and how deep it would penetrate the population, and at what cost? Certainly, there would be an uptick in the number of fresh cases and the number of people admitted to the hospital.

Ideally, you would hope to achieve herd immunity. To achieve herd immunity, you need more than 60% to 70% of the population exposed to the virus and who have developed an immunity.

But that was a long shot, given how long it would take to get 50% to 75% of the population exposed to the virus. We did not even know how much herd immunity would be required for COVID-19 before we could consider it no longer a serious national threat.

It is hoped that, should a vaccine be developed, it will be freely available worldwide with companies willing to forego profit to limit suffering and protect humanity.

In the US, unrest began to erupt before the conclusion of the first wave. Besides facing the COVID-19 medical challenges, Americans were also facing political pressure and the aftermath of the lockdown.

The COVID-19 pandemic had already sunk the American GDP by 3.8%, and close to 30 million Americans were unemployed. These were not speculations, but a reality that had transpired in fewer than 60 days.

In December and January, we were flying around like free birds. In March, we felt like caged animals. How long were we going to endure that, and at what further loss of life, devastation of civilization, and damage to the economy?

This was not an economic diatribe, but a human drama that had far-reaching repercussions in terms of loss of livelihood, loss of family members, domestic violence, untold debts, and loss of earning capacity. It was threatening the very foundation of a sound and prosperous family life.

Nik Nikam
April 3 · 🏷 frontlines update

CORONAVIRUS PANDEMIC WORLD NEWS 4-3-2020

Nik Nikam, MD, MHA. HOUSTON, TEXAS.

AS THE CORONAVIRUS WORLD TURNS ON APRIL 3, 2020

As of April, 3, 2020, the number of cases of coronavirus worldwide stand at 1,015,466 and the total number of deaths at 53,190. The number of cases continue to escalate, and in some places like New York exponentially.
... See More

COVID-19 CORONAVIRUS PANDEMIC

Coronavirus Cases:

1,015,466

Deaths:

53,190

Recovered:

212,229

WORLD / CORONAVIRUS / UNITED STATES

🇺🇸 United States

Coronavirus Cases:

245,066

Deaths:

6,075

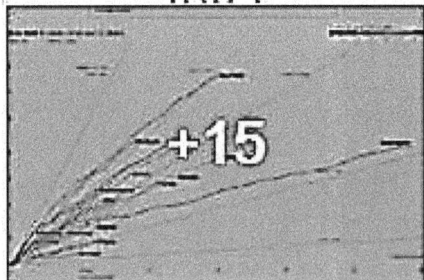

+15

👍❤️😮 Parul R Shah, Tabby Acnp and 65 others   34 Comments

👍 Like   💬 Comment

4   R0 (R NAUGHT) RATE OF VIRAL SPREAD

Most viruses replicate in the host in addition to causing illness, allowing the infected person to spread the virus to the next person, and the next person, and the cycle repeats. In this process, many people may get ill from the virus. You see this phenomenon with the common 'flu.

If you are in close contact with a person with 'flu, there is an extremely high chance you will get the 'flu. Some viruses are so virulent that an infected person can pass it to more than one person, thus doubling or tripling the rate of spread.

The number of people affected by one infectious individual is represented by the 'R0' value (expressed as R Naught).

R0 represents the average number of people infected by one infectious individual. If R0 is greater than 1, the number of infected people will increase exponentially, and an epidemic could follow. If R0 is less than 1, the viral spread is likely to slow or fade out on its own.

R0 alone cannot forecast an outbreak, but it could signal how quickly the infection could spread in the community.

The R0 for SARS-CoV-2 infection was 2.5, significantly higher than the estimates for MERS, but relatively like that of the SARS virus, which caused a deadly global epidemic in 2003.

However, R0 is notoriously complicated to track. It depends not only on the biological characteristics of a virus, which are unknown at the beginning of an epidemic, but also on how often people are exposed to the infected individual. That also provides a partial clue to a workable solution.

This doesn't take into account the asymptomatic carriers, who have never been tested and who could still spread the virus. Until you know the percentage of asymptomatic people who could be spreading the virus, you could be underestimating the R0 value.

About 43% of residents surveyed in the northeastern Italian town of Vo' in February and March tested positive despite having no symptoms. In the US, an estimated 35% of people with COVID-19 infection were asymptomatic.

If you could isolate every infected individual that had the potential to spread the virus, you would, in essence, be reducing the R0 value to below 1. Then, you would have better control over the viral spread, decrease the number of infected people, lessen the demand on the healthcare system, and perhaps cut morbidity and mortality.

Different viruses have unique abilities to infect other people. For example, the common 'flu has an R0 value of 1.3 compared with measles, which has an R0 value of 15 to 18.

It also depends on the incubation period, which is the time it takes to get the symptoms after a person has been exposed to an infectious person. With common 'flu, it could be a couple of days. Whereas, with coronavirus, it takes 1-2 weeks to develop symptoms after exposure to a person with an active infection.

One person with 'flu has the potential to infect 1.3 persons. Those people can then infect another batch of 1.3 people. That may not seem like a substantial number. However, again, don't let numbers deceive you!

One person with measles can infect 15 to 18 people. Each one of those persons can infect 15 to 18 more people. Hence, in a matter of days, thousands of people can get infected through one infected person.

COVID-19, with an R0 of 2.3, would require exposure of 70% of the population to attain herd immunity. The best number we had seen was from New York City, where they demonstrated that 21% of people tested had COVID-19 antibodies. This was a rough estimate of the extent of the immunity and meant that, short of a vaccine, there was no chance for herd immunity. (Figure 4-1)

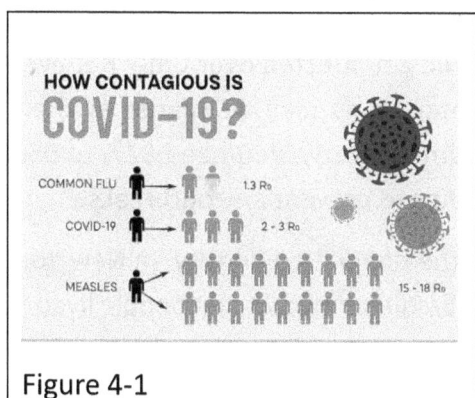

Figure 4-1

Gaining herd immunity by allowing 70% of the population to be infected with coronavirus is ethically and politically unconscionable. However, Sweden tried for herd immunity with little or no success. In addition to having extremely high mortality, Sweden failed to achieve herd immunity. Antibody tests performed in Sweden in August showed that only 14% of the population had antibodies.

The infected person passes on the virus to the next person while talking, coughing, or sneezing, in the form of droplets. This underscores the importance of masks to prevent exposure to those droplets during close contact with infected persons.

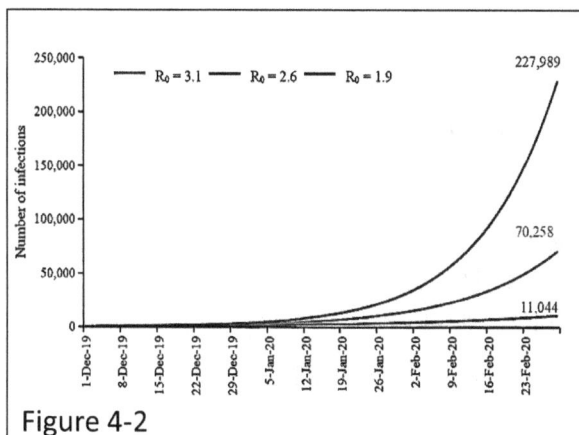

Figure 4-2

It also highlights the importance of a reliable vaccine in case of highly contagious viral infections.

The figure below illustrates how fast the virus could spread and infect so many people over time. A person having a virus with an R0 of 1.9 can spread the disease to 11,044 in fewer than 85 days. A person with a virus such as COVID-19, which has an R0 value of 2-3m, can spread the virus to 70,258 people. A person having a virus with an R0 value of 3.1 can spread the disease to 227,989 people. (Figure 4-2)

Could you guess the number of people one person with measles (with an R0 of 15 to 18) could spread to in the same duration? With no vaccination, roughly 95 percent of the population would get infected over time. But even with that serious infection, 5% of the population may be spared as there would be no one to catch the disease from. Similarly, you need 92% to 95% of the population vaccinated against mumps to prevent any outbreaks.

Another dimension to consider was the population density. In New York City, the population density was 27,755.25/square mile. When people lived in small crowded places, with common elevators, hallways, and central air conditioning, the spread was inevitable.

The case doubling rate, a derivative which is based on the R0 value of the COVID-19 and local circumstances, was closely followed and compared to that of other countries.

As increased measures such as quarantine, contact tracing, face masks, and self-mitigation rules are put in place, the number of infected people and new cases decreases. This secondary rate of infection is referred to as Re.

Using this sound scientific concept, German Chancellor Angela Merkel announced in April that they could bring the Re from 1.1 to 0.7, which reduced their active cases and kept their mortality rate relatively low.

Similarly, countries like Hong Kong, South Korea, Taiwan, New Zealand, and Japan, which instituted self-mitigation guidelines, not only contained the viral spread but also were able to keep their active case numbers low while keeping their businesses open.

After Texas lifted its lockdown toward the end of June, there was a sudden surge of new cases—as high as 24% of the people tested. That created an enormous surge in the number of active cases and hospital admissions, overburdening the ICU capacity. It was also associated with an increase in mortality. That proved the R0 or Re could swing to both extremes, depending on how well steps to prevent the COVID-19 spread were implemented.

CASE DOUBLING RATES

During the COVID-19 peak pandemic phase, many countries' doubling rates ranged from three days to 18 days. The sure way to reduce the doubling rate was to slow the spread of the coronavirus by identifying the active cases and isolating them. But, when the healthcare workers and the healthcare leaders were battling an active onslaught of sick people, it was challenging to focus on isolating active cases and following their contacts. (Figure 4-3)

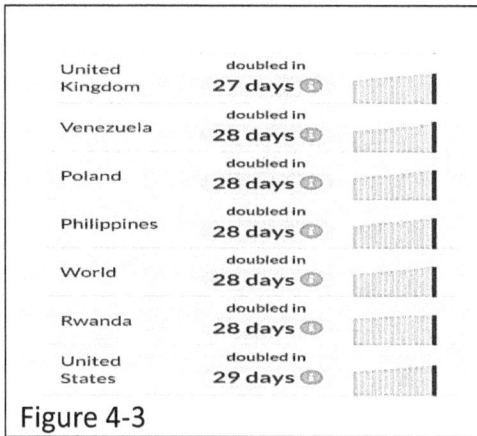

United Kingdom	doubled in 27 days
Venezuela	doubled in 28 days
Poland	doubled in 28 days
Philippines	doubled in 28 days
World	doubled in 28 days
Rwanda	doubled in 28 days
United States	doubled in 29 days

Figure 4-3

Ref:

https://www.sciencealert.com/why-herd-immunity-will-not-save-us-from-the-covid-19-pandemic

https://www.the-scientist.com/features/why-r0-is-problematic-for-predicting-covid-19-spread-67690

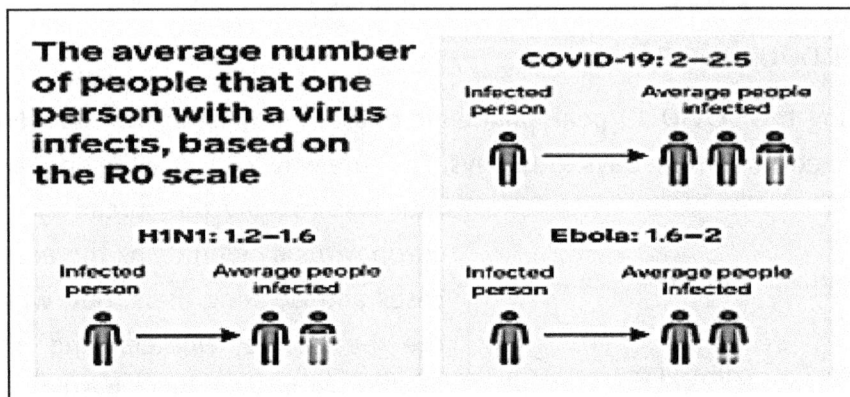

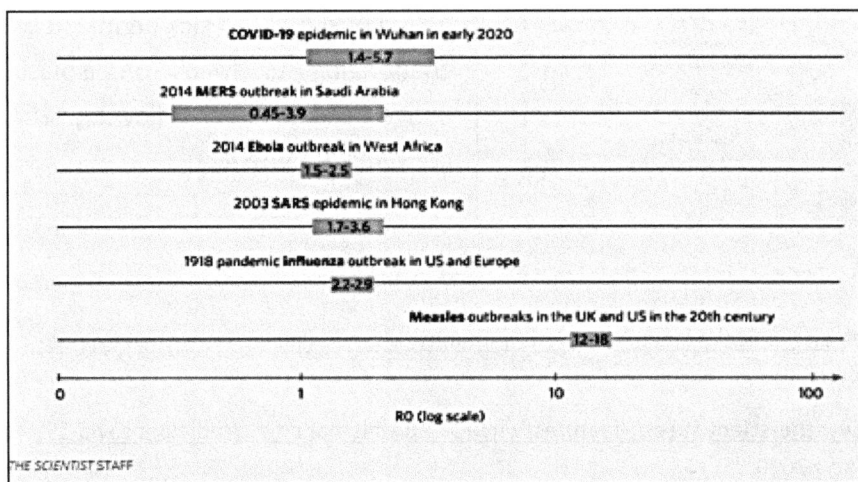

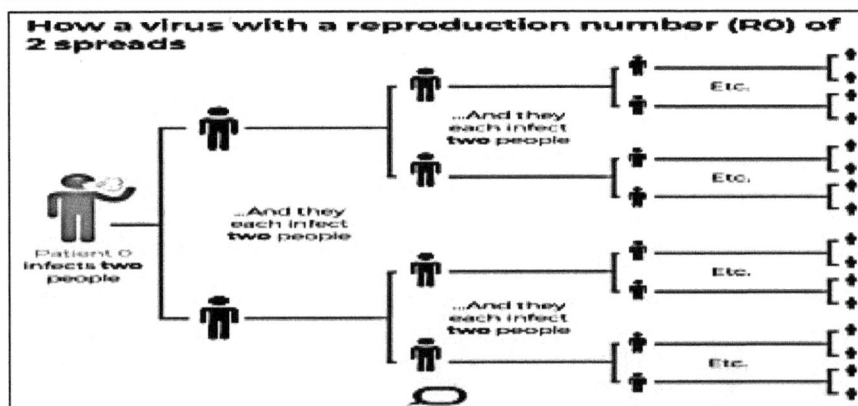

Figure 4-4

5   MASKS AND MYTHS!

There was an intense debate about wearing masks to prevent the spread of coronavirus. At the start of the COVID-19 pandemic, there was little scientific evidence to support mask-wearing. What evidence we had involved experiences with diseases such as 'flu, SARS, and MERS. In Hong Kong, people exposed to SARS were familiar with the importance of masks in curbing viral spread; hence, they had more faith in masks in the early stages of the pandemic and had greater success in reducing COVID-19 spread.

The first purpose of masks was to prevent a person with COVID-19 infection from spreading the virus to others. The second purpose of masks was to protect people from getting COVID-19 infection from others. That is all the information they needed for safety.

You might question what a person with a known COVID-19 infection was doing in public in the first place. Shouldn't that person be in quarantine and away from the public? Yes! But it was not possible to know who was infected early on in the pandemic as it was not possible to test the entire population, and there were many asymptomatic cases, which was another variable that made things more complicated. Several studies found that 35% to 45% of cases were asymptomatic.

So, if these people had no tests to detect the coronavirus, they might spread the virus to more vulnerable people while remaining asymptomatic themselves. As there were many asymptomatic people in the general population, it was reasonable to assume you could get the virus from anyone you might encounter. Therefore, it made sense for everyone to wear masks.

However, the WHO and the CDC promoted conflicting recommendations from time to time. Contrary to their fleeting recommendations, as medical professionals, we felt that masks were the best means by which to minimize the chances of catching the virus.

The CDC put out a handy tutorial on its website on how to make masks using household materials like old T-shirts, bandanas, coffee filters, or old hair ties. This was a good Do-It-Yourself home project for those who wanted to kill time and serve humanity at the same time.

Research had shown if 80% of the population were to wear masks, the viral spread could be slowed by 60%. This was better than the best treatment we had at the time for COVID-19 patients.

A vaccine would take 12 to 18 months to become available to the public. That signaled that you could be in for a lengthy, bumpy ride with the masks, and you might as well adopt them as a lifesaver.

HOW WAS COVID-19 TRANSMITTED FROM PERSON TO PERSON?

The coronavirus measures one micron in size. A droplet is about 5 to 10 microns. COVID-19 transmission was primarily through those large droplets, which carried many clusters of coronaviruses. When someone sneezed or coughed, they expelled several hundred droplets. Hence, without a protective mask, the chances of inhaling those droplets and catching COVID-19 were high.

Late in August 2020, the experts reported the virus also could be spread by aerosols, which contain much smaller particles, and cloth masks also could block those aerosol particles.

Viral droplets can also settle on objects and surfaces, and COVID-19 can be active on surfaces from four to 24 hours, depending on the surface. If you encountered infected surfaces with your hands and then rubbed your hands-on your face, you ran the risk of catching the virus.

A common scenario was a computer keyboard used by several people. Similarly, door handles, furniture, elevator buttons, or handrails could harbor the virus for hours or days.

The COVID-19 virus attaches to receptors in the nose, which act as a gateway to the body. Scientists found that specific cells in the nose had elevated levels of proteins that the COVID-19 virus used to get into our cells. Failing to wear a mask covering your nose would not protect others should you sneeze, releasing droplets into the air and potentially infecting other people.

A research study from Hong Kong showed wearing a mask lowered the viral transmission rate through airborne particles or respiratory droplets by 50% to 75%. Denying or defying such epidemiological data only exposed people to unnecessary COVID-19 infection.

WHO SHOULD BE WEARING A MASK?

It was of paramount importance for a person with an infection to wear a mask to prevent the droplets from landing on other people's faces. That was more important than a person wearing an N95 mask to prevent getting infected by the coronavirus.

You wanted to prevent those infectious droplets from reaching healthy people. As it was difficult to know who had the infection and who did not, it was a safe compromise to make sure everyone had a mask, even if it was made of cloth. That applied even to those who already had an infection, to show a token of solidarity to curb this virus.

WHY SHOULD EVERYONE WEAR A MASK?

In a fast-moving pandemic, everything is new, constantly changing, and fleeting. Unexpected turns keep everyone guessing, and constant change leads to confusion and irrational decisions that can cost lives and cause devastation to families and the economy. Hence, the most practical decision you can make in a pandemic is to simply wear a mask. If everyone wears a mask early in a pandemic, the spread is limited, saving many lives and allowing a return to work, preserving the economy and community.

STURGIS COVID-19 STUDY

Between August 7 and August 16, 2020, more than 500,000 motorcycle enthusiasts gathered for an annual rally in Sturgis, South Dakota. It was a large gathering with little, if any, observation of self-mitigation guidelines.

In the coming weeks, the counties that contributed to the highest number of attendees experienced a 7.0 to 12.5% increase in COVID-19 cases relative to the counties that did not participate. The economic cost was close to $12.2 billion.

That underscored the importance of continuing self-mitigation principles until a vaccine is available.

WHICH MASK FOR WHAT PURPOSE?

Cloth masks were effective in preventing large virus droplets from spreading in the air and exposing nearby people. Austria noted a 90% decline in infection rate two weeks after the implementation of mandatory masks.

If you want to prevent getting an infection, you need something like an N95 mask that prevents the droplets from entering your nose, air passages, and parts of your face. Cloth masks block both the droplets and aerosol spread. N95 masks block 95% of the small particles.

DOS AND DON'TS WITH YOUR MASK!

Note that the outside of your mask might contain droplets; hence, it could be infectious.

I saw so many people constantly adjusting their masks, potentially negating the usefulness of the mask due to cross-contamination from droplets on the outside of the mask.

When you put on a mask, avoid touching the outside surface. Use the elastic bands to secure the mask firmly over your face. After that, you shouldn't be meddling with your mask constantly.

It was difficult to wear the mask for extended periods. If you don't have a mask that snugs your face, you might be tempted to constantly touching and adjusting the nose part, so it is vital that you have a mask that fits properly.

Some people dropped the mask below the nose, which negated the entire purpose. Some people took their mask off and put it on their desk, exposing them to whatever was on the outside of the mask. Some pulled the mask down to talk to people, risking contaminating their hands with the mask and passing on a potential infection to the people they were talking with. This highlights the importance of educating people on proper mask use to avoid mask contamination and actions that render mask-wearing useless.

WHAT WAS THE ROLE OF A FACE SHIELD?

The face shield protected not only air passages but also eyes and skin. That prevented the droplets from landing on your eyes or uncovered parts of your face. If you touched those areas, you might get infected. Healthcare workers dealing with COVID-19 patients wore face shields as they faced a bigger viral load. Sanitizing sprays or solutions easily cleaned the face shields.

NOT ALL MASKS WERE CREATED EQUAL!

The CDC recommended people to wear any kind of mask in public places and at workplaces, especially if they can't maintain a 6-foot distance.

Masks comprising two layers of batik quilting fabric or cotton backed with flannel were effective in blocking droplets.

Homemade masks did not prevent a healthy person from getting the virus from a person who had a COVID-19 infection who was not wearing a mask, but a homemade mask could prevent the infected person from spreading those droplets to healthy people. You want to control the virus at the source, and in that regard, it helped to have a cloth mask over wearing nothing.

Homemade cloth face masks should be wide enough to cover your nose and chin and snug enough to not create gaps. That tight fit prevents the droplet particles from traveling through the air and escaping through the gaps. Micro-filters inserted between the fabric of the homemade masks block the virus droplets from getting into your airways.

If you use cloth face masks outside the home, carefully remove them without touching the outside layer. Wash the masks immediately in hot/warm soap and water, and when dried, they will be ready for reuse.

Ultraviolet light can also be used to clean the mask, as it was shown to kill the virus in minutes. The hospitals used large UV light robots that cleaned COVID-19 patient rooms. Keeping the masks outside on a sunny day with a temperature of more than 95 degrees sterilizes the mask in 20 minutes. New techniques and UV devices for sterilizing masks and other personal items such as keys and cell phones rapidly entered the market. Also, if you put the mask aside for ten days, any virus on that surface should be dead by that time.

EFFICACY OF MASK USE

Japan instituted masks for everyone. Their death rates were 10% of the US numbers. Their death rates may also have been lower due to other factors such as ethnicity, no strikes or protests, the number of tests done, reporting criteria, etc.

Japan's densely packed population was 1/5 of the US population. It had no lockdown, had still-active subways, and many businesses remained open, including karaoke bars. Their citizens took part in social distancing where they could. They did not heavily invest in contact tracing.

Similarly, whenever we saw images of Wuhan, we saw thousands of people walking with masks.

People from the far east, such as Asia, India, or China, might have had natural immunity from exposure to many communicable diseases, along with BCG vaccination and climate.

The Czech Republic made face masks mandatory in mid-March. Within a week, they had ten million homemade masks. They had one of the lowest mortalities in the EU.

Similarly, Hong Kong mandated mask-wearing even after the epidemic had substantially reduced, which may explain why they did not experience a second wave. Beijing, China's largest city, detected several new cases of COVID-19 in the Xinfidi market, which provided 80% of the produce and meat to the people in the city.

On May 25, George Floyd, an African American man choked to death by a white officer on the streets of Minneapolis, Minnesota, while several onlookers pleaded for his life. That sparked nationwide protests against police brutality. People took to the streets in major cities in large numbers. Most people taking part in the protests were not wearing masks or maintaining social distancing.

Those large gatherings could have increased the risk of coronavirus spread and might have been partly responsible for the second spike in the middle of June. Additionally, when policymakers opened Florida beaches after the lockdown period, most people who attended had no masks.

After states eased lockdown, many people became complacent with the self-mitigation rules. They attributed these factors as plausible causes for the second spike seen in Texas, Florida, California, and 20 other states.

COVID-19 PATIENTS AT HOME

If a person infected with COVID-19 lived with others, they wore a surgical or cloth mask at home to reduce the chance of spreading the virus within the household, and other household members wore masks to avoid contracting the virus from the infected person. However, if you live in a small area, it might be difficult to contain the virus. Often, multiple household members tested positive.

CAN'T WE GET THE INFECTION AND CALL IT DONE?

Some people said, "let me get the virus and get over it like the common 'flu, so I can get back to my routine." Not so with the coronavirus. You did not know who would develop lung complications that could be fatal.

When more than 50% of the US experienced the second spike in the number of new cases, the admissions of people between the ages of 20 and 40 went up, with many of them in the ICUs.

Therefore, the risk of intentionally contracting the virus is too high, and wearing a simple mask is a much better option. *If you don't enjoy wearing a mask, you won't enjoy being on the ventilator with a tube in your airway.*

CAN PROLONGED MASK USE LEAD TO CARBON DIOXIDE POISONING?

Some felt wearing a mask lead to a build-up of carbon dioxide (hypercapnia), causing symptoms such as lightheadedness, fatigue, and headache. There was no evidence that carbon dioxide level build-up was high enough to cause symptoms. When people were working alone at their desks, they did not need a mask. In most cases, people might wear the mask for an hour or two at a stretch, which was not long enough to cause any carbon dioxide related symptoms.

The mask must cover the nose because the COVID-19 attaches to the receptors in the nose, which is the gateway to the airways and lungs. Likewise, if you sneeze with your nose uncovered, you could spread the virus to people around you.

Did you know the surgeons wear masks for eight to ten hours during surgery? They don't complain of symptoms related to carbon dioxide. Hence, there is no scientific evidence that wearing a mask for a few hours is harmful.

Masks don't eliminate the need for other self-mitigation techniques such as social distancing, washing hands, and staying at home if possible. All these measures were complementary to minimize the exposure to the virus and thus slow the spread of the COVID-19 spread.

MASKS FOR HEALTHCARE WORKERS DEALING WITH COVID-19 PATIENTS

Healthcare workers dealing with COVID-19 patients need professional N95 respirators. Professionals usually fit the N95 masks to individual build, so they fit snugly over their faces. That stops most of the droplets from getting into the nose and air passages. Also, some healthcare workers use face shields to protect the areas that are uncovered by masks, such as the eyes, nose, mouth, and skin. Health officials reserve these shields for first responders and those in the front lines dealing closely with COVID-19 patients. They also wore gowns to protect their clothes from contamination with viral particles. They removed the gowns after seeing patients.

POLITICAL AND SOCIAL CONCERNS

We must deal with political activists on both sides. According to activists, a blanket mask mandate places a limit on individual liberty and even one's right to free speech. In the initial stages of the COVID-19 pandemic, there was a tremendous shortage of surgical and N95 masks, even for the frontline workers and the healthcare personnel.

LESSONS FOR THE FUTURE PANDEMICS

One important lesson for the professionals and political leaders dealing with future epidemics is to make sure to have a stockpile of masks and personal protective equipment at the earliest news of an epidemic, anywhere in the world. No masks and no personal protective gear meant a free spread of the virus-like a wild forest fire!

Ref:

cdc.gov

who.int

https://cheps.sdsu.edu/docs/Contagion_Externality_Sturgis_Motorcycle_Rally_9-5-20_Dave_et_al.pdf

https://www.washingtonpost.com/health/2020/09/08/worst-case-scenerios-sturgis-rally-may-be-linked-266000-coronavirus-cases-study-says/

Mask Type	Standards	Filtration Effectiveness		
Single-Use Face Mask	China: YY/T0969	Open-Data Tests Smart Air SmartAirFilters.com 3.0 Microns: ≥95% 0.1 Microns: ✘		
Surgical Mask	China: YY 0469	3.0 Microns: ≥95% 0.1 Microns: ≥30%		
	USA: ASTM F2100	Level 1 3.0 Microns: ≥95% 0.1 Microns: ≥95%	Level 2 3.0 Microns: ≥98% 0.1 Microns: ≥98%	Level 3 3.0 Microns: ≥98% 0.1 Microns: ≥98%
	Europe: EN 14683	Type I 3.0 Microns: ≥95% 0.1 Microns: ✘	Type II 3.0 Microns: ≥98% 0.1 Microns: ✘	Type III 3.0 Microns: ≥98% 0.1 Microns: ✘
Respirator Mask	USA: NIOSH (42 CFR 84) China: GB2626	N95 / KN95 0.3 Microns: ≥95%	N99 / KN99 0.3 Microns ≥99%	N100 / KN100 0.3 Microns ≥99.97%
	Europe: EN 149:2001	FFP1 0.3 Microns: ≥80%	FFP2 0.3 Microns: ≥94%	FFP3 0.3 Microns: 99%

3.0 Microns: Bacteria Filtration Efficiency standard (BFE).
0.1 Microns: Particle Filtration Efficiency standard (PFE).
0.3 Microns: Used to represent the most-penetrating particle size (MPPS), which is the most difficult size particle to capture.
✘: No requirements.

smartairfilters.com

Filtration efficacy for various Materials

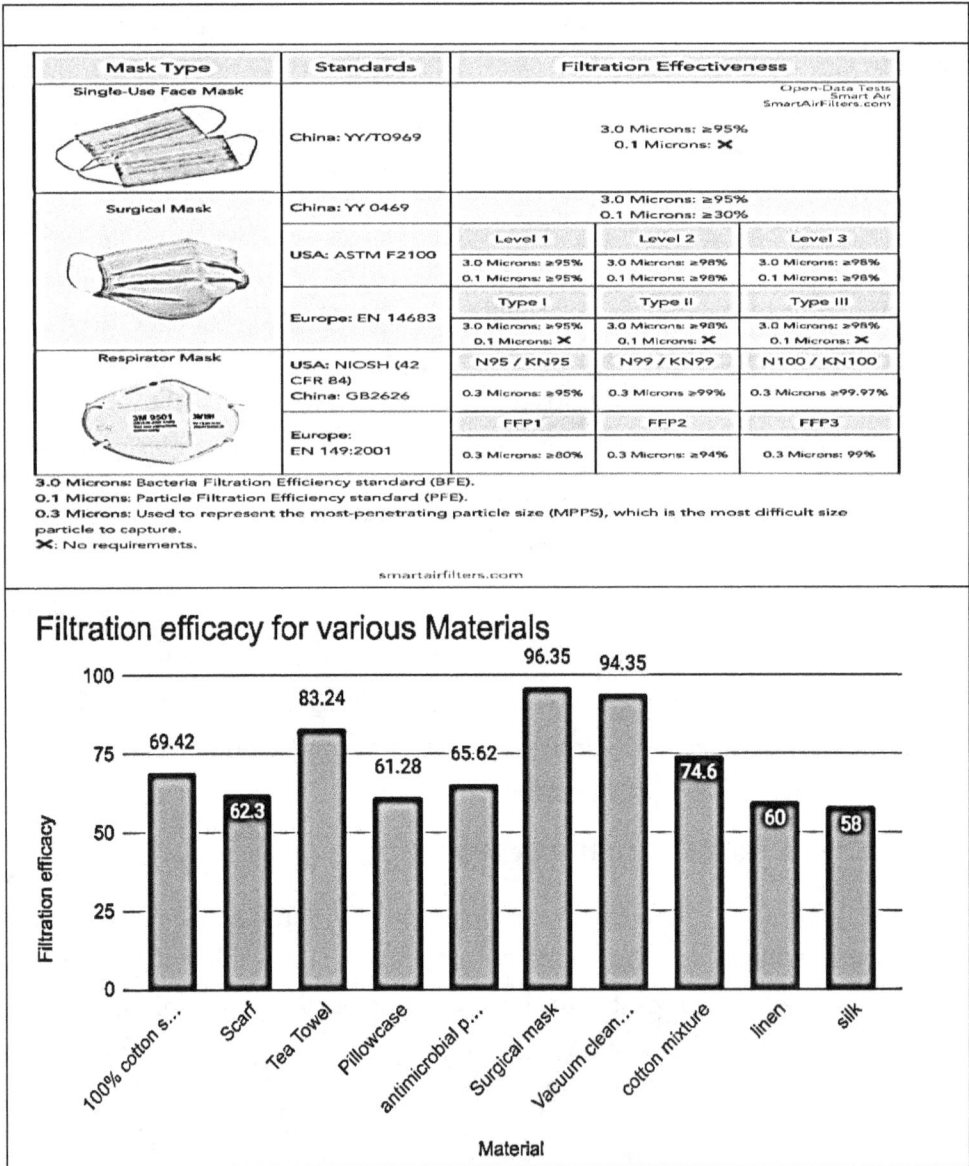

Bar chart — Filtration efficacy (y-axis) vs Material (x-axis):
- 100% cotton s...: 69.42
- Scarf: 62.3
- Tea Towel: 83.24
- Pillowcase: 61.28
- antimicrobial p...: 65.62
- Surgical mask: 96.35
- Vacuum clean...: 94.35
- cotton mixture: 74.6
- linen: 60
- silk: 58

6   PATHOGENESIS

SARS-CoV-2 targets the nasal and bronchial epithelial cells and pneumocytes. The spike protein has an affinity to the angiotensin-converting enzyme 2 (ACE2) receptor.

The type 2 transmembrane serine protease in the pneumocytes (alveolar epithelial cells) promotes viral uptake by cleaving ACE2 and activating the SARS-CoV-2 S protein, which mediates viral entry into host epithelial and endothelial cells.

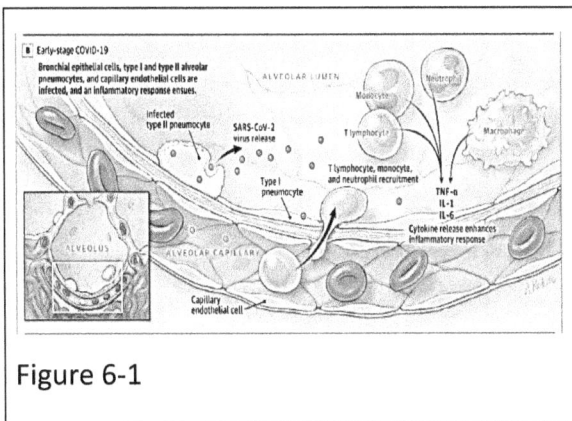

Figure 6-1

The SARS-CoV-2 also attacks the T lymphocytes and causes severe leukopenia. The virus sets off both the innate and adaptive immune responses, impairing lymphopoieses, and enhancing lymphocyte apoptosis. (Figure 6-1)

ACE2 receptor upregulation by ACE inhibitors and angiotensin receptor-blocking drugs were thought to increase susceptibility to the COVID-19 infection; however, several studies showed no increased risk among people taking those medicines.

In the later stages of infection, the viral replicates compromise epithelial and endothelial barrier integrity. That triggers an influx of monocytes and neutrophils.

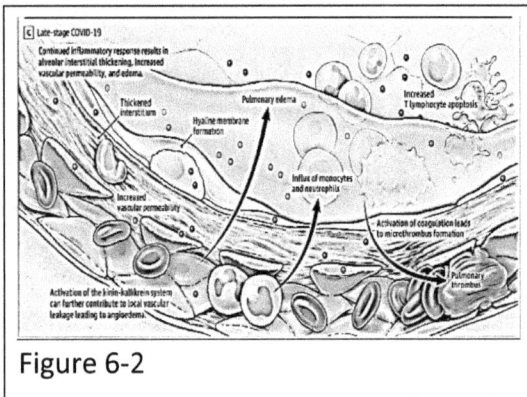

Figure 6-2

Autopsy studies showed diffuse thickening of the alveolar walls with mononuclear cells and macrophages infiltrating airspaces. Interstitial inflammatory mononuclear infiltrates and edema lead to ground-glass opacities on computed tomographic (CT) images. Thick gelatinous hyaline membrane filled the alveolar spaces leading to pulmonary edema, compatible with the early phases of acute respiratory distress syndrome (ARDS). Bradykinin-dependent lung angioedema might also have contributed to lung infiltrates. (Figure 6-2)

Those cellular changes lead to endothelial barrier interruption, impaired alveolar-capillary oxygen transmission, and reduced oxygen diffusion capacity. In advanced cases, there was a fulminant activation and consumption of coagulation factors. In a report from Wuhan, China, 71% of 183 patients who died from COVID-19 met the criteria for diffuse intravascular coagulation. The inflamed alveolar tissues and pulmonary endothelial cells might have caused microthrombi, contributing to a higher incidence of thrombotic events such as deep venous thrombosis, pulmonary embolism, or arterial thrombotic complications. In advanced critical cases, the dysregulation of host responses resulted in the viral infection leading to multiorgan failure.

Endothelial cells were essential contributors to the initiation and propagation of severe COVID-19, involving multiple systems, as the endothelial cells (EC) lined every organ's blood vessels. ACE2 receptors are located not only on the alveolar epithelial cells but also on endothelial cells that line the vascular system. Disruption of endothelial cells leads to inflammation, massive microthrombi, reduction of blood supply, and eventual micro-infarcts.

An acute respiratory distress syndrome (ARDS) and an acute hypoxic respiratory failure was the leading cause of mortality in patients with COVID-19. The endothelial cells, along with the surrounding mural cells (pericytes), maintain the vascular integrity and barrier function. Emerging evidence suggested the involvement of the endothelial cells initiated and propagated ARDS by altering vessel barrier integrity, promoting a pro-coagulative state, inflammation (endotheliitis), and inflammation-mediated cellular and immunohumoral changes.

As many as 30% of hospitalized patients developed severe disease with progressive lung involvement, in part due to an overactive inflammatory response. Vascular barrier breaches lead to tissue edema (fluid in the alveoli), endotheliitis, activation and dissemination of intravascular coagulation, and deregulated inflammatory cell infiltration.

Reduced ACE2 activity indirectly activated the kallikrein–bradykinin pathway and increased vascular permeability. Activated neutrophils, recruited to pulmonary ECs, produced histotoxic mediators, including reactive oxygen species (ROS). Immune cells, inflammatory cytokines, and vasoactive molecules lead to enhanced EC contractility disrupting inter-endothelial junctions.

Finally, the cytokines IL-1β and TNF activated glucuronidases that degraded the glycocalyx, upregulated hyaluronic acid synthase 2, leading to increased deposition of hyaluronic acid in the extracellular matrix and promoting fluid retention in the alveoli.

On the left, the normal interface between the alveolar space and endothelial cells is depicted; the right-side highlights pathophysiological features of COVID-19 in the lung, including loss of vascular integrity, activation of the coagulation pathway, and inflammation. (ROS, reactive oxygen species; S1PR1, sphingosine 1 phosphate receptor 1; VWF Von Willebrand factor).

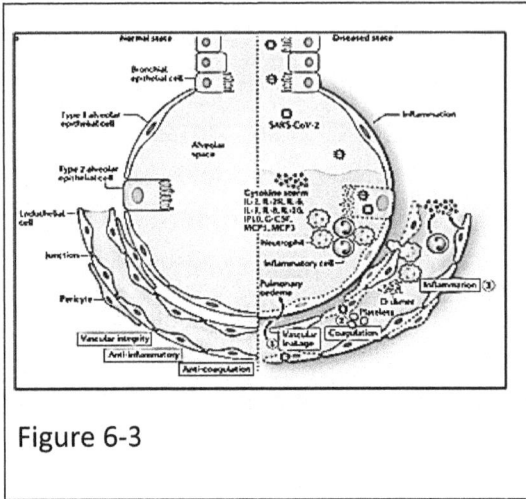

Figure 6-3

Coagulation cascade: Endothelial activation, dysfunction, and death lead to exposure of the thrombogenic basement membrane. IL-1B and TNF initiated coagulation by expressing P-selectin, Von Willebrand factor, and fibrinogen, to which platelets banded. The EC released tropic cytokines further augmented platelet production. (Figure 6-3)

Platelets released VEGF, triggered ECs, and activated pericytes to upregulate the tissue factor, the prime activator of the coagulation cascade. (Figure 6-4)

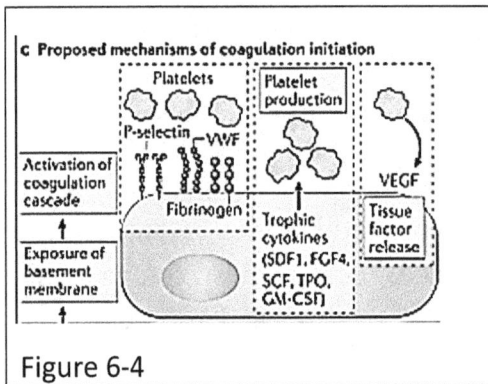

Figure 6-4

In response, the body-mounted countermeasures to dissolve fibrin-rich blood clots, explaining why elevated levels of fibrin breakdown products (D-dimers) were predictors of poor patient outcomes. Angiogenesis counteracted the ischemia. However, EC cells lining the new vessels faced the same challenges.

Cytokine Storm: Many patients with severe COVID-19 showed high levels of cytokines, amplifying the destructive process by leading to further EC dysfunction, DIC, inflammation, and vasodilation of the pulmonary capillary bed, alveolar dysfunction, ARDS with hypoxic respiratory failure, and ultimately multiorgan failure and death. (Figure 6-5)

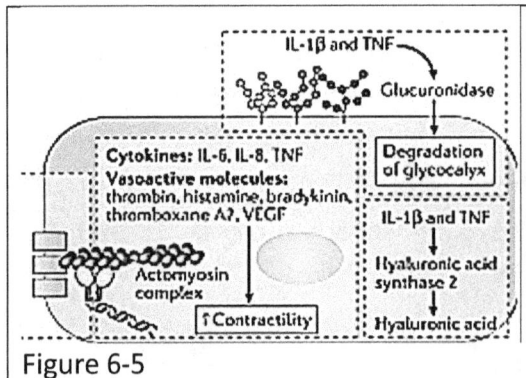

Figure 6-5

ECs promoted inflammation by expressing leukocyte adhesion molecules, facilitating the accumulation and extravasation of leukocytes, including neutrophils, which enhanced tissue damage.

Denudation of the pulmonary vasculature leads to activation of the complement system, promoting the accumulation of neutrophils and pro-inflammatory monocytes that enhanced the cytokine storm. They based this theory on the observation that, during influenza infection, pulmonary ECs induced an amplification loop involving virus-infected pulmonary epithelial and interferon-producing cells. ECs were the gatekeepers of this immune response, as modulation of the sphingosine 1 phosphate receptor 1 (S1PR1) in pulmonary ECs dampens the cytokine storm in influenza infection. That raised the question as to whether pulmonary ECs had a similar function in the COVID-19 cytokine storm and whether S1PR1 could represent a therapeutic target.

Another unexplained observation was the excessive lymphopenia in severely ill patients with COVID-19 and whether this was related to the recruitment of lymphocytes away from the blood by activated lung ECs. Normalization of the vascular wall through metabolic interventions could be

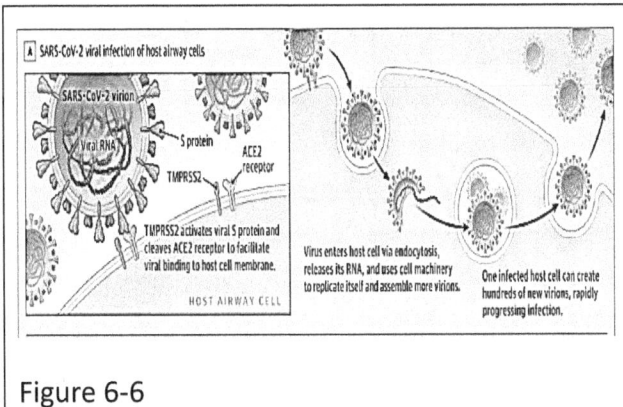

Figure 6-6

an additional route of intervention. (Figure 6-2)

Ref:

https://jamanetwork.com/journals/jama/fullarticle/2768391

https://www.nature.com/articles/s41577-020-0343-0

https://www.roche.com/investors/updates/inv-update-2020-07-29.htm?

https://rebelem.com/covid-19-acute-lung-injury-a-proposed-model-v1-0-via-farid-jalali-md/

https://www.nature.com/articles/s41577-020-0343-0?fbclid=IwAR04ZLLq9CZ578P6XntDd-81CmNpDL-oFNZel8r-CodKN4b3XamJ4i06kmw

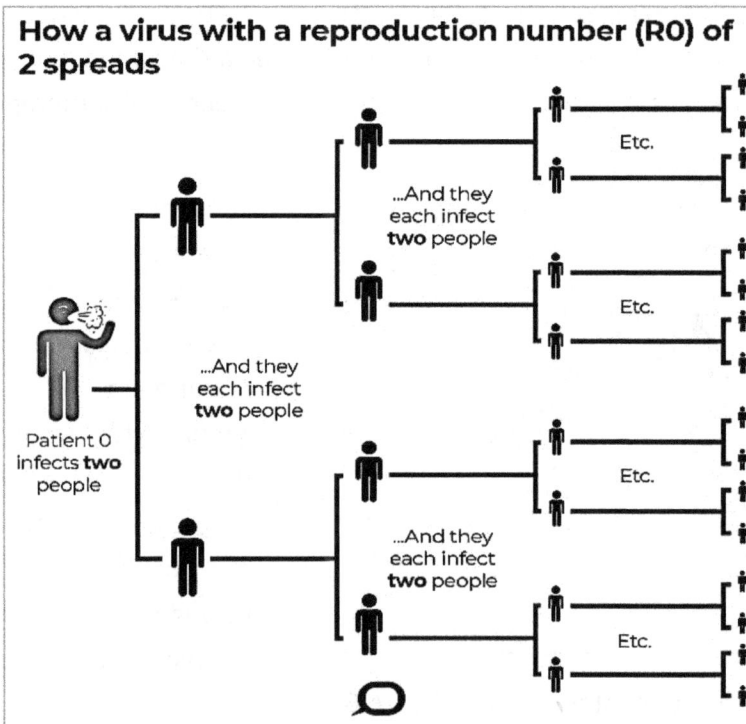

How a virus with a reproduction number (R0) of 2 spreads

7 HYPERIMMUNE GLOBULINS AND MONOCLONAL ANTIBODIES

INTRODUCTION

Hyperimmune globulin is a concentrated version of the immune body from convalescent plasma from an infected patient. One advantage of using this antibody approach for COVID-19 is there is no need for blood typing before administration.

Hyperimmune globulin from the plasma of COVID-19 patients provides passive immunity when delivered to seriously ill patients during acute illness to enhance the body's ability to fight the virus, reduce viral multiplication, and improve clinical outcomes. The antibody protection may last from a week to a few months. On the other hand, vaccines stimulate the body's immune system to produce antibodies and provide long-term immunity.

HISTORICAL

Concentrated antibody treatment in animals in the 1890s showed that when serum from diphtheria-infected animals was injected into other animals early in their infection course, it prevented them from developing severe diphtheria illness. This formed the basis for large-scale production of diphtheria and tetanus antibodies from horses, cows, and sheep. Other noted antibody treatments include rabies, pneumococcal pneumonia, measles, and polio. Most modern-day antibodies have been of human origin to prevent serum sickness from animal antibodies. This strategy enabled us to use immunoglobulins against hepatitis A & B, and Varicella-Zoster.

The body's immune system attacks foreign substances by making large numbers of protein antibodies that stick to a specific antigen protein. Once the antibodies find and attach to the antigen, other parts of the immune system destroy the pathogen containing the antigen. (Figure 7-1)

IgM anti-S and anti-N immunoglobulins appear approximately one week after infection and continue to increase over two weeks, while IgG appears by the third week.

Antibody titer is determined by serial dilution to calculate the dilution required to maintain enough antibodies for an effect to be achieved.

A serial dilution method is a technique used to estimate the concentration (the amount of antibody) of an unknown sample by counting the number of antibodies from serial dilutions of the sample and then back-track the measured counts to the unknown concentration. Thus, an antibody detected at 1:1000 dilution has a much higher concentration than the one detected only at 1:100 dilution.

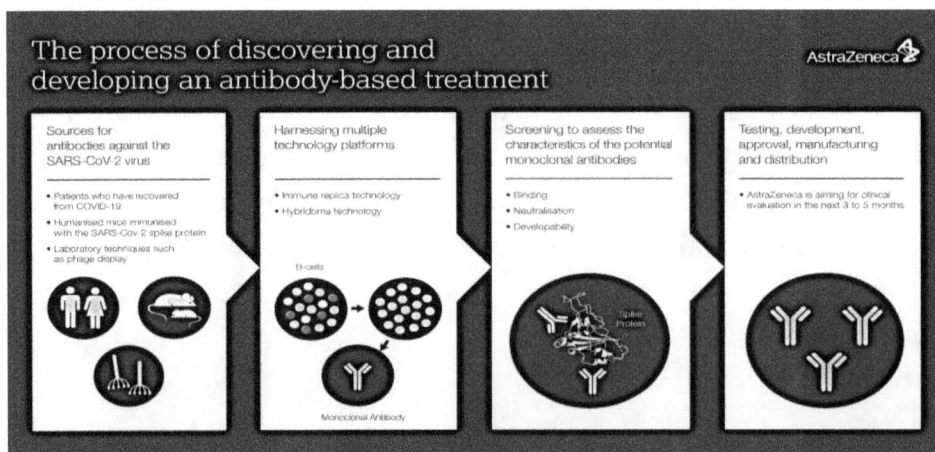

Figure 7-1

HYPERIMMUNE GLOBULINS IN COVID-19

Research laboratories produced hyperimmune globulins from large quantities of plasma obtained from COVID-19 patients. One potential disadvantage included the cost of collection, processing, and pathogen inactivation steps for large quantities of plasma. Hyperimmune globulins were a reasonable alternative to convalescent plasma and perhaps carried a lower risk of adverse events. Typically, hyperimmune globulin is given as a single dose.

However, the efficacy of hyperimmune globulin in COVID-19 patients was unknown. It was assumed that if antibodies in plasma provided immunity, the hyperimmune globulins with comparable properties against the COVID-19 would have similar benefits.

There were several advantages to using hyperimmune globulins:

- Less contamination with pathogens
- Lower volume
- More concentrated or higher titers of antibodies
- Lower transfusion reactions and other adverse events
- Potential for intramuscular administration
- Ease of storage, shipping, and reaching remote regions of active outbreaks

MONOCLONAL ANTIBODIES

Monoclonal antibodies are synthetic proteins that act like human antibodies. They are proteins and could cause a rare allergic reaction.

Hypersensitivity reactions to monoclonal antibodies can be classic type I (mast cell-mediated, perhaps IgE dependent) reactions, cytokine release reactions, or type IV cell-mediated reactions.

Such reactions also can be effectively limited or prevented with appropriate premeditations, intravenous fluids, and dose or frequency adjustment of the monoclonal antibody.

Type IV cell-mediated reactions, such as erythema multiforme, Stevens-Johnson syndrome, toxic epidermal necrolysis, drug reaction with eosinophilia and systemic symptoms, and other blistering reactions are absolute contraindications to re-exposure to the implicated agent.

Some monoclonal antibodies act as immunotherapies as they make the immune system respond better and able to attack pathogens more effectively.

COVID-19 ANTIGENIC TARGETS

Spike protein. The spike protein is a transmembrane surface glycoprotein that binds to the angiotensin-converting enzyme 2 (ACE2) receptor on respiratory epithelial cells and gastrointestinal cells, mediating

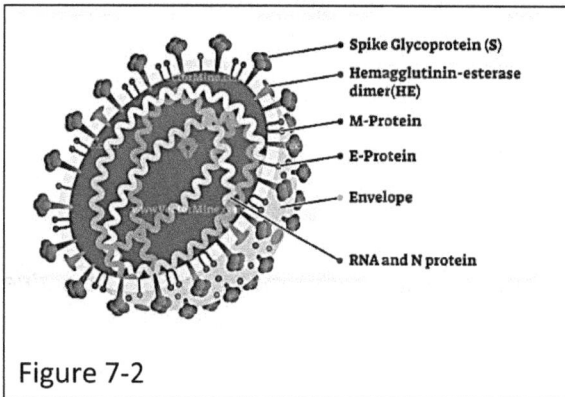

Figure 7-2

viral entry. Antibodies against the spike protein target of COVID-19, the receptor-binding domain (RBD), might block viral entry into the respiratory epithelium, thus reducing the viral multiplication or duration of infection. (Figure 7-2)

Labels in figure: Spike Glycoprotein (S); Hemagglutinin-esterase dimer(HE); M-Protein; E-Protein; Envelope; RNA and N protein

Nucleocapsid protein. The nucleocapsid protein interacts with the viral nucleic acid (RNA) and contributes to the assembly of functional virions. Antibodies to nucleocapsid protein are developed along with anti-S, but their role in recovery from the infection is ill-defined.

Assays that rely on antibody binding, such as enzyme-linked immunosorbent assay (ELISA)-type assays, have been developed for antibodies against both S and N.

TYPES OF ASSAYS

An assay is a test used to determine the quality or the quantity of an antibody or an immunoglobulin. Antibody titers are measured functionally or serologically:

Bioassays: Bioassays assess the effect of plasma on virus viability (viral neutralization). This involves live viruses; hence, it requires special precautions to avoid laboratory contamination or infection of personnel.

Neutralization assays: Here, expression of individual viral proteins (but not a live virus) is assessed in cell culture treated with plasma with or without antibodies. Cell culture is a method used to cultivate, propagate, and grow a large number of cells in a dish.

Serological assays: Most laboratories use a serologic assay that measures antibody binding to a target antigen (an enzyme-linked immunosorbent [ELISA]-type assay). Some assays use chemiluminescence as the readout as it is more sensitive.

Correlations between live virus, pseudo-virus neutralization, and ELISA titers are limited.

OPTIMAL TITER

In the US, the FDA recommended a titer of 1:160. It stated a titer of 1:80 might be acceptable if plasma with higher titers were not available. However, clinical trials testing convalescent plasma had used different methods and different titer thresholds, ranging from 1:160 to 1:1000. Studies were underway to assess the effectiveness of larger doses.

Testing the donor during a pre-donation visit was more practical; they could defer those who did not have a sufficient titer. It was unknown how much time could elapse between donor testing and the reduction of their titers. The titers needed reassessment every two weeks.

The strength and durability of the immune response was a crucial factor in determining how long vaccines would be effective and how often people might need a booster dose.

HYPERIMMUNE GLOBULIN AND ANTIBODY STUDIES

The Mount Sinai Health System partnered with BioSolutions and ImmunoTek Bio Centers to develop and test a COVID-19 hyperimmune globulin product on health care providers at elevated risk for COVID-19 infection and other high-risk populations. This study was funded with a $34.6 million Department of Defense grant.

South Korea tested a hyperimmune globulin antibody from the plasma of recovered COVID-19 patients.

Grifols is the worldwide leader in the production and sale of immunoglobulins.

GigaGen used a recombinant hyperimmune drug class for COVID-19 patients, which had 100-fold higher potency than convalescent serum. They could produce millions of doses using serum from a few donors who had developed antibodies to the SARS-CoV-2 virus.

Recombinant convalescent serum provides purity, consistency, and potency, along with proven efficacy, diversity, and polyvalence.

Another form of antibody treatment was called the monoclonal antibodies, known as 'mAbs.' MAbs typically have two or three antibodies that are easy to manufacture, which are safe and nontoxic.

According to a BioRxiv study, the antibody levels in donor serum can vary by as much as 1000-fold.

On August 25, British pharmaceutical giant AstraZeneca began an antibody-drug trial for the treatment and prevention of COVID-19.

Three potential scenarios where antibodies could be useful include patients with low-level responses to vaccines (e.g., the elderly), as a therapeutic early in the disease, or in people who don't want to take the vaccine.

Other competitors in the antibody race include Regeneron, which is already in human clinical trials, along with Amgen, AstraZeneca, GlaxoSmithKline, and Eli Lilly.

Ily Lilly received Emergency Use (EUA) Authorization for its mAbs Bamlanivimab on November 13, 2020.

Ref:

https://www.aha.org/news/headline/2020-07-09-mount-sinai-joins-effort-develop-covid-19-hyperimmune-globulin-product

https://www.biospectrumasia.com/news/39/16436/gc-pharma-files-ind-for-hyperimmune-globulin-based-therapy-for-covid-19.html

https://www.ncbiotech.org/news/grifols-produces-first-batches-potential-covid-19-therapeutic-testing

https://ccpp19.org/healthcare_providers/hyperimmune_globulin/index.html

https://www.biospace.com/article/gigagen-publishes-data-on-new-recombinant-hyperimmune-drug-class-as-covid-19-therapy/

https://www.fiercebiotech.com/research/gigagen-s-polyclonal-antibody-against-covid-19-outperforms-plasma-lab-tests

https://www.uptodate.com/contents/coronavirus-disease-2019-covid-19-convalescent-plasma-and-hyperimmune-globulin#:~:text=HYPERIMMUNE%20GLOBULIN%20As%20noted%20above,early%20in%20the%20disease%20course.

https://www.cancer.org/treatment/treatments-and-side-effects/treatment-types/immunotherapy/monoclonal-antibodies.html

https://www.cslbehring.com/vita/2020/convalescent-plasma-and-hyperimmune-globulin

https://www.biorxiv.org/content/10.1101/2020.06.17.153486v1.full

https://www.cnbc.com/2020/10/05/trumps-use-of-regenerons-experimental-coronavirus-treatment-creates-very-tough-situation-ceo-says.html

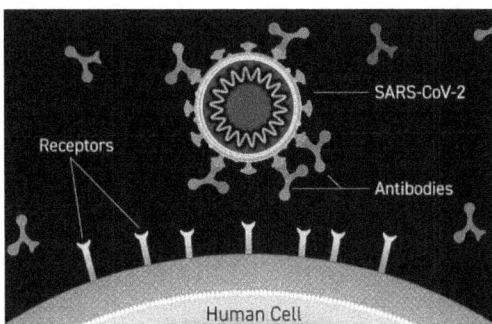

Nik Nikam
June 12 at 6:01 PM · public health

TALE OF TWO MIDDLE EASTERN COUNTRIES -CORONAVIRUS
PANDEMIC - TURKEY V. IRAN!
A PREVIEW OF A POSSIBLE SECOND WAVE?
Nik Nikam, MD. MHA. HOUSTON, TX
Turkey and Iran are very rich middle east countries with rich cultural
heritage, oil and natural resources. The Turkey has a population 84 million
which is comparable to the Iranian population of 83 million. Both courtiers
are advanced medical services. Turkey is the world's 17th-largest country by
population.

...
See More

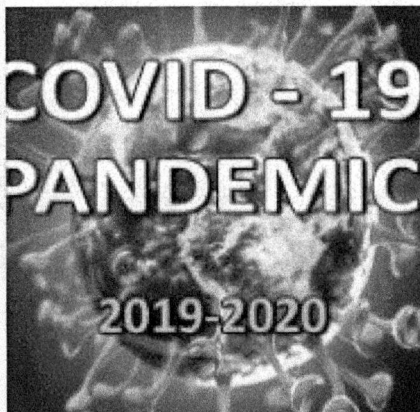

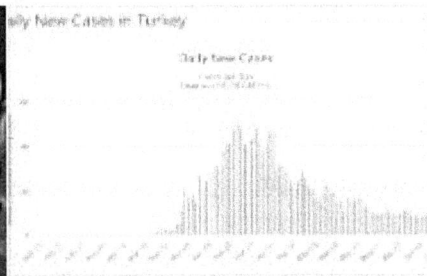

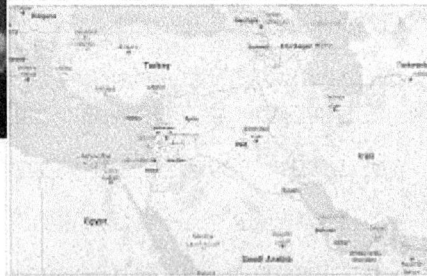

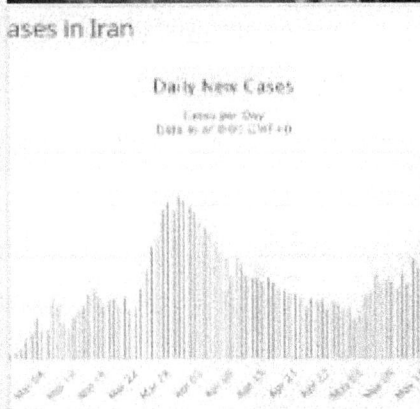

	DEATHS	TEST
14	4729 +5	2,4151
27	8425	4.348,8

Ramandeep Kahlon and 25 others   35 Comments

8   CLINICAL PRESENTATION

WHAT IS CORONAVIRUS?

Coronaviruses are large, single-stranded RNA viruses found in humans and animals such as dogs, cats, bats, chicken, and cattle. SARS-CoV-2 is the third coronavirus that has spread globally and caused severe respiratory disease in humans in the past two decades.

Bats are a natural reservoir for coronaviruses. The coronavirus is constantly mutating to survive in the host. The genome of SARS-CoV-2 is similar to that of other bat coronaviruses, as well as those of pangolins.

When human beings encounter such animals, they are exposed to various genetic forms of the coronavirus. However, the viruses must mutate and adapt to the new host to be able to survive, multiply, and spread. SARS-CoV-2 is different from the other coronaviruses in that it has a spike protein that binds well with another protein on the outside of human cells called ACE2. This enables the virus to hook into and infect human cells.

The other coronaviruses included SARS-CoV, which caused the 2020 Severe Acute Respiratory Syndrome (SARS) pandemic and originated in

Foshan, China, and the coronavirus that caused Middle East Respiratory Syndrome (MERS), which originated from the Arabian Peninsula in 2012.

SARS-CoV-2, which causes COVID-19, has a diameter of 60-140 nm and has distinctive spike proteins, ranging from 9-12 nm, making the virions look like a solar corona. (Figure 8-1)

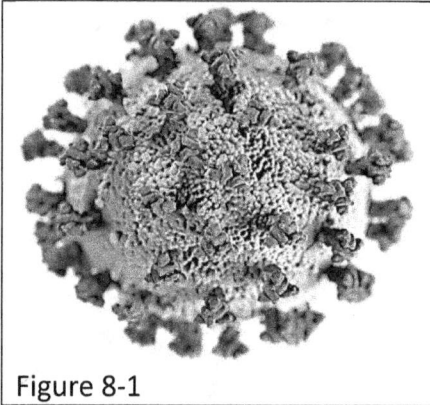

Figure 8-1

TRANSMISSION

Exposure to an infected person within six feet for at least 15 minutes, or briefer exposures to symptomatic patients, increases the risk for transmission. Pre-symptomatic carriers accounted for more than 50% of transmission.

COVID-19 spreads primarily through respiratory droplets during close face-to-face contact. Both symptomatic and asymptomatic patients can spread the virus.

One or two viruses might not make you sick. Scientists estimate you need more than 1000 viruses before you could get sick with the virus.

Being outdoors also helped to increase social distance. The wind spread the droplets, so you were less likely to get a large viral dose.

Aerosols are microscopic particles that can remain airborne for hours and potentially carry pathogens for several feet. Aerosol particles are smaller than five micrometers (0.005 millimeters) in diameter. Larger droplets expelled by sneezing or coughing fall to the ground or other surfaces quickly, while aerosols are suspended in the air for minutes to hours. How long a virus can remain airborne depends on the size of the particles containing it.

Touching an object exposed to the virus could be another mode of transmission. Higher viral loads persisted for 3-4 days on impermeable surfaces, such as stainless steel and plastic than on permeable surfaces, such as cardboard or wood. That might explain how the virus could contaminate objects such as doorknobs, rails, handles, cutlery, keyboards, or clothing.

However, the predominant mode of transmission of COVID-19 is through droplets spreading during face-to-face contact. Infection was more frequently noted in clusters of people who had been in close contact, such as attending church gatherings or singing in a choir. Hospital rooms were also noted to have widespread viral contamination. Hence, hospitals had to decontaminate a patient room occupied by COVID-19 patients immediately after they left.

The viral load in a patient's air passages peaked 2-3 days before symptom onset. Pre-symptomatic individuals accounted for 48% to 62% of viral transmission. People could be asymptomatic during the first week and still could spread the virus to many other people. Even though viral nucleic acid chains could be found in throat swabs for up to six weeks after the onset of illness, the viral cultures become negative eight days after symptom onset.

Mothers infected with COVID-19 during the third trimester did not seem to transmit the virus to the neonates.

The Center for Disease Control and Prevention recommends isolation for at least ten days after symptom onset and three days after the improvement of symptoms.

CLINICAL PRESENTATION

The mean incubation period for COVID-19 from exposure to symptom onset infection is approximately five days (range: 2-14 days). Approximately 97.5% of individuals develop symptoms within 11.5 days of infection. Most people (98%) had symptoms by day 12. Over 80% of symptomatic patients experienced mild symptoms, and around 45% of infected individuals had no symptoms. The median interval from symptom onset to hospitalization was 7 (3-9) days. The median age of hospitalized patients was between 47 and 73 years, with a male preponderance (60%).

However, when the southern states such as Florida, Texas, and California experienced the second surge of new cases, the majority age group had shifted to 20-40-year-olds. This was largely due to younger individuals neglecting the self-mitigation guidelines and heading to the bars, beaches, and clubs with no regard for mask mandates or social dispensing.

The COVID-19 spectrum included asymptomatic carriers to those with a fulminant disease with sepsis, respiratory failure, coagulation disorders, and cytokine storm. Common symptoms included fever, dry cough, shortness of breath, fatigue, nausea, vomiting, and severe muscle aches.

Approximately 5%–10% of patients with COVID-19 required hospitalization. About 20% of hospitalized patients developed more serious symptoms, requiring intensive care—over 5% of hospitalized patients required oxygen. (Figure 8-1)

Symtpoms	Percentage
Fever	70%-90%
Dry cough	60%-86%
Shortness of breath	53%-80%
Fatigue	38%
Myalgias	15%-44%
Nausea/vomitting	10%-39%
Headche	25%
Running nose	7%
Loss of smell	3%

Figure 8-2

EXAMINATION

Most Covid-19 patients presented with a mild fever. The physical findings were minimal. Later, some patients developed crackles or coarse noises at the lung base, suggesting lung involvement. Cytokine storm was a serious late complication that led to death in many incidences. (Figure 8-2)

CLINICAL COURSE

Stage 1: COVID-19 started with mild symptoms such as fever, dry cough, myalgia, and nausea. The predominant laboratory abnormality was leukopenia. Most people did well in a home setting with fluids, pain medicines, and rest. (Figure 8-3)

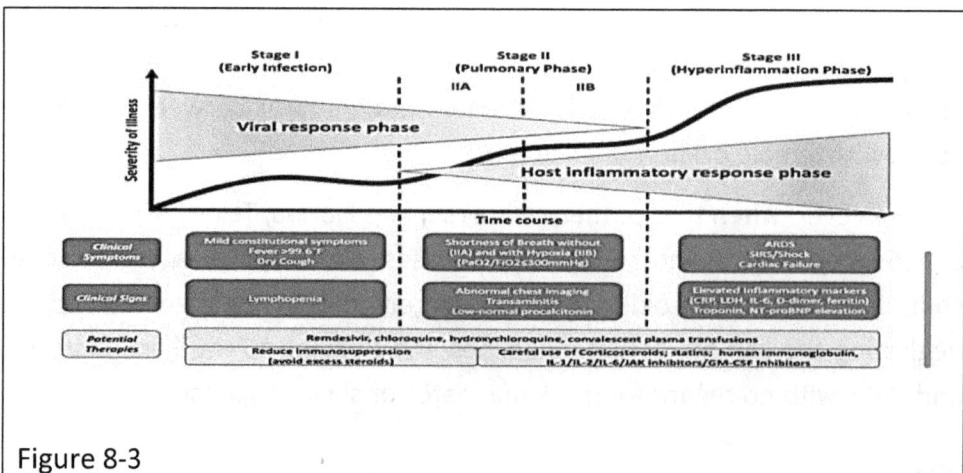

Figure 8-3

Stage 2: Patients progressing to this stage developed shortness of breath and hypoxia (low oxygen levels). These patients had abnormal chest X-rays and CT scans, elevated enzymes, pro-calcitonin, and inflammatory markers.

Stage 3: This was the more fulminant phase where the hyperimmune response to the viral infection led to adult respiratory distress syndrome, shock, and cardiac failure with widespread thromboembolic complications. This was associated with the elevation of markers, including Sed. Rate, CRP, ferritin, Interleukin, LDH, D-dimer, troponin, ProBNP, etc.

LABORATORY FINDINGS

Reverse transcription-polymerase chain reaction (RT-PCR) was used to diagnose active COVID-19 infection.

RT-PCR is a laboratory technique combining reverse transcription of RNA into DNA and amplifying the specific DNA targets using polymerase chain reaction. Combined RT-PCR is gene expression analysis and quantification of viral RNA in a clinical setting.

Twenty percent of individuals might have a false negative test. Factors contributing to false-negative tests included inadequate specimens, time from exposure, and specimen sources. The timing of the RT-PCR test was important. A positive test result was found in 33% of individuals tested four days after exposure, 62% at the symptom onset, and 80% three days after symptom onset.

Several serological tests were available to aid in the diagnosis and measurement of the body's response (antibodies) to infection and novel vaccines. The serological assays included point-of-care assays and high-volume enzyme immunoassays. However, test performance, accuracy, and validity were variable. Additionally, the presence of antibodies did not necessarily imply immunity, as not all antibodies detected were found to be neutralizing, which is necessary for an antibody to be effective. How long antibody protection would last was unclear. IgM antibodies appeared within five days after symptom onset and peaked around 2–3 weeks. IgG antibodies appeared 14 days after symptom onset.

In some cases, blood count showed leukopenia or reduction in the leukocytes and elevated lactic dehydrogenase. A Chinese study of 2874 patients showed alterations in serum C-reactive protein, lactate dehydrogenase, and albumin. The most common hematological abnormalities were lymphopenia, coagulopathy, modest prolongation of prothrombin times, thrombocytopenia, and elevated D-dimer. However, most of these laboratory characteristics were nonspecific and are common in pneumonia. (Figure 8-4)

Abnormal Labs.	Results
Lymphopenia	Present
Sed Rate	Elevated
C reactive protein	Elevated
Ferritin	Elevated
Tumor necrosis factor- α,	Elevated
II-1 to IL-6	Elevated
Prolonged PT/INR	Elevated
Thrombocytopenia	Low
Elevated D-Dimer	Elevated

Figure 8-4

The D-dimer and, to a lesser extent, lymphopenia, had the largest prognostic associations.

Radiographic findings of COVID-19 infection included bilateral, lower-lobe infiltrates on chest radiographic imaging. A chest CT was more reliable in assessing the extent of lung damage, characterized by bilateral peripheral fluffy pulmonary infiltrates involving the lower-lobes with ground-glass opacities. (Figure 8-5)

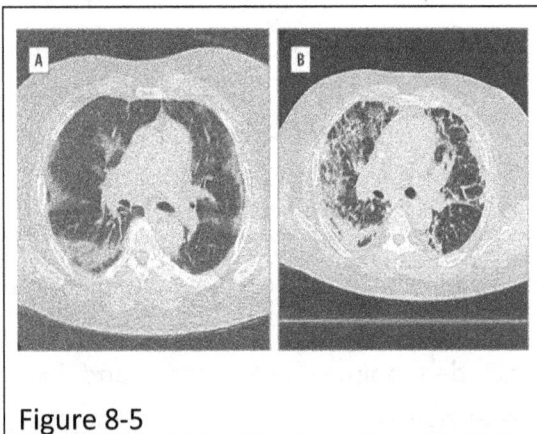

Figure 8-5

The other CT finding included ill-defined margins, air bronchogram, smooth or irregular interlobular or septal thickening, and thickening of the adjacent pleura.

The rapid evolution of abnormalities typically occurred in the first two weeks after symptom onset, after which the changes subsided gradually. The chest CT findings were nonspecific and overlapped with other infections, providing limited diagnostic value in COVID-19 patients.

COMPLICATIONS

Frequent complications included pneumonia, acute respiratory distress syndrome (ARDS), and myocarditis. There was widespread damage to capillary endothelial cells leading the microthrombi in the heart, liver, kidney, brain, and the peripheral arterial system.

Rare, but lethal, complications included cytokine storm, macrophage activation syndrome, sepsis, and multiorgan failure. Physicians noted a loss of taste and smell in 64%–80% of patients.

COMPLICATIONS	%
Pneumonia	75%
ARDS	15%
Cardiac injury	7%-17%
Thromboembolic events	10%-25%
Kidney injury	9%
Neurological complications	8%
Stroke	6%
Septic shock	6%

Figure 8-6

Approximately 2%–5% of COVID-19 patients were younger than 18 years. Children had milder symptoms limited to the upper respiratory tract and rarely required hospitalization. Only 2-3% of hospitalized children developed severe lung complications. (Figure 8-6)

TREATMENT

Patients were admitted to the hospital if they had multiple risk factors, symptoms, and abnormal tests. Asymptomatic stable patients were advised to quarantine at home and watch for any alarming symptoms like high fever, shortness of breath, or persistent cough.

Most patients were admitted for observation due to fever or shortness of breath. They received supportive care, analgesics, oxygen when needed, and IV fluid for dehydration. Their oxygen level and respiratory levels were closely observed for signs of pulmonary decompensation.

Many countries and professional societies, including the National Institutes of Health (NIH), established evidence-based guideline initiatives. They constantly updated their guidelines based on the latest information and developments.

Over 75% of hospitalized patients with COVID-19 required supplemental oxygen. Patients who were unresponsive to conventional oxygen therapy received heated high-flow nasal cannula oxygen. For patients needing invasive mechanical ventilation, lung-protective ventilation with low tidal volumes (4-8 mL/kg, predicted body weight), and plateau pressure less than 30 mm Hg was used. Additional measures to improve oxygenation included prone positioning, a higher positive end-expiratory pressure strategy, and short-term neuromuscular blockade with cisatracurium or other muscle relaxants.

Even though COVID-19 patients showed lung complications, they still benefited from lung-protective ventilation. Early intubation allowed time for a controlled intubation process with less contamination. However, many patients tolerated hypoxemia well in the absence of respiratory distress and did not require mechanical ventilation.

At the time of writing, there were no concrete recommendations regarding earlier versus later intubation. Earlier intubation could expose some patients to additional complications.

Around 8% of COVID-19 hospitalized patients had bacterial or fungal infections, and 72% of hospitalized patients received broad-spectrum antibiotics.

Targeting the Virus and the Host Response

Classes of drugs tried in the management of COVID-19 patients:

- ✓ Antivirals (Remdesivir, Favipiravir),
- ✓ Anti-inflammatory agents (Dexamethasone, statins),
- ✓ Immunomodulatory therapies (Tocilizumab, Sarilumab, Anakinra, Ruxolitinib),

- ✓ Anticoagulants (heparin, Lovenox) and antifibrotics (tyrosine kinase inhibitors).
- ✓ Antibodies (convalescent plasma, hyperimmune immunoglobulins),

Different treatment modalities had different efficacies at various stages of illness and in dissimilar manifestations of the disease. Viral inhibitors were most effective early in infection, while, in hospitalized patients, immunomodulatory agents were most useful to prevent disease progression, and anticoagulants prevented thromboembolic complications in some patients. Most hospitalized patients received thromboembolic prophylaxis with subcutaneous low molecular weight heparin. However, they could still develop clots due to damaged blood vessel cells.

Hydroxychloroquine. Hydroxychloroquine was thought to have beneficial immunomodulatory effects, reduce viral entry into alveolar epithelial cells, and block viral replication. After more than 200 clinical trials, there was much debate around the efficacy and safety of hydroxychloroquine in the treatment of COVID-19. The clinical trials did not demonstrate an obvious benefit, except for the Henry Ford Hospital study involving 2540 patients which showed a reduction in mortality. That study is covered elsewhere in this book. Adverse effects included QT prolongation and increased cardiac arrhythmias. There are a few ongoing randomized trials to study outcomes in COVID-19 patients receiving hydroxychloroquine.

Doctors tried most antivirals that had been tested previously against influenza, HIV, Ebola, and SARS/MERS in COVID-19 patients and found no statistically significant improvement in mortality, reduction in ICU admissions, or need for ventilators.

Protease inhibitor lopinavir-ritonavir, which disrupted viral replication *in vitro*, failed to show benefit when compared with standard care in a randomized, controlled, open-label trial of 199 hospitalized adult patients with severe COVID-19 infections.

RNA-dependent RNA polymerase inhibitors such as ribavirin, favipiravir, and remdesivir were also assessed in COVID-19 patients. The remdesivir study, involving 1063 adult patients receiving intravenous remdesivir for up to ten days, reduced the hospital recovery time from 15 days to 11 days, even though there was no improvement in any other endpoints like mortality or need for ventilators. A separate study involving 397 patients showed a 10-day treatment with remdesivir was no more effective than five days of treatment.

Scientists used convalescent plasma during the 1918 Spanish 'Flu pandemic. In an initial study involving five COVID-19 patients, a substantial improvement in symptoms, oxygen level, organ failure, viral load, antibody titers, and a need for extracorporeal membrane oxygenation support was noted. However, an incomplete study from China involving 103 patients showed no benefit. Lack of enrollment ended the study before completion.

More recently, the Mayo Clinic reported on more than 70,000 convalescent plasma transfusions and found a 35% reduction in complications when the plasma was given within 48 hours after the diagnosis of COVID-19 infection. Please see a separate chapter on Plasma transfusion for COVID-19 patients.

Hyperimmune globulin and monoclonal antibodies derived from convalescent plasma from COVID-19 patients were in a clinical trial at the time of writing.

Alternative therapeutic strategies included monoclonal antibodies directed against key inflammatory mediators such as interferon-gamma, interleukin 1, interleukin 6, and complement factor 5a to prevent organ damage.

A Phase 3 trial using the interleukin 6 inhibitors tocilizumab and sarilumab showed no benefit in hospital-mortality, length of hospital stay, need for ventilators, among others.

Tyrosine kinase inhibitors, such as imatinib, were evaluated for their potential to prevent pulmonary vascular leakage in individuals with COVID-19, with no significant positive outcomes.

The Randomized Evaluation of COVID-19 Therapy (RECOVERY) trial, where 2104 patients received 6 mg daily of dexamethasone for up to ten days versus 4321 received usual care, found that dexamethasone reduced 28-day all-cause mortality (21.6% vs. 24.6%; age-adjusted rate ratio, 0.83 [95% CI,

0.74-0.92]; P < 0.001). This study is discussed in detail in a subsequent chapter.

There was a disproportionate prevalence of COVID-19 and associated elevated mortality rates among Black patients. A study from Louisiana showed that Black patients accounted for 77% of hospitalization for COVID-19 infection and 71% in-hospital mortality, even though only 31% of the total population is Black. Minority groups often lived in densely populated communities or housing, depended on public transportation, worked in crowded places, and had a higher prevalence of chronic health conditions than white individuals. People with low socioeconomic status also had more co-morbid conditions and extremely limited access to affordable healthcare.

PROGNOSIS

Overall hospital mortality from COVID-19 was 15%–20%, rising to 40% among ICU patients. It was lowest among children, ages 1–7, and highest among those over the age of 80 years. Black and Hispanic people had higher mortality than white individuals. Patients with a history of heart disease, diabetes, or obesity had higher morbidity and mortality; (Figure 8-7)

GROUPS	MORTALITY
Overall inpatient	14–20%
ICU Patients	>40%
Ages <40 years	>5%
Ages 70–79 years	>35%
Ages 80–89 years	>60%
Outpatient all	0.02%

Figure 8-7

However, patients with chronic lung disease did not have higher mortality compared with those without lung disease. (Figure 8-8)

COVID-19 mortality in the US varied with age and comorbid conditions. The mortality was 0.3 deaths per 1000 cases among those aged 5 to 17. It jumped to 304.9 per 1000 cases in patients 85 years or older. The mortality was as high as 40% in the intensive care unit patients.

PRE-EXISTING CONDITION	DEATH RATE confirmed cases	DEATH RATE all cases
Cardiovascular disease	13.2%	10.5%
Diabetes	9.2%	7.3%
Chronic respiratory disease	8.0%	6.3%
Hypertension	8.4%	6.0%
Cancer	7.6%	5.6%
no pre-existing conditions		0.9%

Figure 8-8

COVID-19 had an alarming potential to cause severe long-term illness and disability. Some patients experienced lingering weakness, fatigue, or shortness of breath. However, most people recovered from the loss of taste and smell 2–4 weeks after recovery.

Multiple organ system damages in some patients led to multiorgan failure. Some patients developed severe lung scarring, leading to long-term loss of function. The blood clotting abnormalities accounted for strokes and kidney failure in previously healthy patients. COVID-19 also caused myocardial inflammation in some patients.

A vaccine might be a definitive answer for a pandemic with the potential to persist for several years to avoid these serious medical complications. This is covered in depth in the chapter entitled Vaccines.

LESSONS FOR THE FUTURE:

A multi-prong approach to control any such future outbreaks should include:

- Self-mitigation steps like masks, physical distancing, personal hygiene, protective equipment at the beginning of the pandemic, and not when the infection had spread to too many people.

- Case detection and isolation; contact identification and tracking; and public education
- Selective school or workplace closure, disinfection
- Regulatory actions like lockdown, masks, mandates, closure of crowded places, restriction of domestic and international travel, etc.

Basic public health interventions have not changed since the 1918 Spanish 'Flu pandemic, which includes home quarantine of infected people, restriction of gatherings, and implementing social distancing along with other self-mitigation steps.

Ref:

https://jamanetwork.com/journals/jama/fullarticle/2768391

https://www.nature.com/articles/s41577-020-0343-0?fbclid=IwAR04ZLLq9CZ578P6XntDd-81CmNpDL-oFNZeI8r-CodKN4b3XamJ4i06kmw

https://jamanetwork.com/journals/jama/fullarticle/2768391

https://www.nature.com/articles/s41577-020-0343-0

https://www.roche.com/investors/updates/inv-update-2020-07-29.htm?

https://rebelem.com/covid-19-acute-lung-injury-a-proposed-model-v1-0-via-farid-jalali-md/

https://www.smithsonianmag.com/science-nature/what-scientists-know-about-airborne-transmission-new-coronavirus-180975547/

CDC.gov

Coronavirus.gov

Who.int

COVID-19 PANDEMIC 2019-2020

Nik Nikam
March 15 · How can I help?

Open letter to Mr. VP Mike Pence and President Donald J. Trump

from Nik Nikam, MD, MHA., DTM and the majority of physicians, dentists, and medical professionals in this group.

Dear Mr. VP Mike Pence and Mr. President Donald J. Trump,

We, the physicians in America, feel it is high time to declare a complete lock down of the entire country, as they did in Wuhan, Italy, and South Korea.

Please take a look at the world Coronavirus statistics as of March 15, 2020.

There is a clear distinction between the countries that have a very low rate of new cases reported, as compared to the US, UK and other European countries.

Countries like Singapore, Japan, South Korea, and Hong Kong had the lowest rate of new cases reported during the last 10 day period.

The number of new cases in South Korea has already leveled off for the past 7 days.

Mr. Vice President and Mr. President the writing is on the wall, or should I say on the picture.

A picture is worth a million words. It may be worth tens of thousand of lives.

As you can see the US new cases is on a steep takeoff, like a military jet. We cannot afford to let this happen.

1. It will skyrocket the number of new cases

2. It will increase the number of infected people from this deadly disease

3. It will tax the healthcare system, lead to exhaustion and more medical professional causalities.

4. It will have a profound impact on our battered economy

5. We will be left behind the rest of the world in overcoming this pandemic.

We should be leading the world in the efforts to confront this Coronavirus, bring an end to this colossal loss of human lives, and prevent a devastating blow to the world economy and the draining of our healthcare systems.

In addition, 20% to 30% of the people are left with significant long term lung problems. We don't need this.

So, we physicians and medical professionals, who work in the trenches along side deathly sick patients with this coronavirus infection, see the human carnage and we can help these people. Mr. President, with your help and power, we can bring an end to this senseless loss of lives due to inaction.

We urge VP Mike Pence and you, Mr. President Donald J. Trump, to take immediate action and institute total lock down so the medical professionals can begin to help those who are already infected and in the hospitals.

We love our medical profession and are willing to sacrifice our own lives to save fellow human beings. However, we don't need to let innocent people die for nothing.

Mr. President Donald J. Trump, you have the presidential power to save this nation, protect people from getting infected, and from dying for nothing.

Country by country: how coronavirus case trajectories compare

Britta Ostermeyer, Zara Khan and 3.1K others   715 Comments

68

9 TREATMENT OPTIONS

Besides supportive treatment, antibiotics for secondary infections, and/or respiratory support, many antiviral agents have been tried with mixed results. Remdesivir showed a reduction in hospital length of stay with no mortality benefit. Thousands of articles were published on the controversial drug hydroxychloroquine with or without zinc and with or without azithromycin with conflicting reports on mortality, safety, and clinical outcomes.

I have reviewed important drugs and studies and highlighted the challenges in performing any studies during a pandemic in which the disease is readily spreading, and the treatment options and our knowledge of the disease is rapidly evolving

COVID-19 causes diffuse lung damage. It was hypothesized that glucocorticoids might modify inflammation-mediated lung injury and reduce progression to respiratory failure and death. However, the small scale non-randomized glucocorticoid studies were inconclusive, and their role remains unclear. Few drug or treatment studies demonstrated any decrease in mortality, though the Remdesivir study showed a reduction in the length of stay in COVID-19 patients.

The pathophysiological features of severe COVID-19 revealed an acute pneumonic process with extensive radiologic opacity. An autopsy showed diffuse alveolar damage, inflammatory infiltrates, and microvascular thrombosis. They also had markedly elevated levels of inflammatory markers, including C-reactive protein, ferritin, interleukin-1, and interleukin-6.

The current COVID-19 pandemic caused by SARS-CoV-2 prompted investigators worldwide to search for an effective antiviral treatment, including several antiviral drugs such as ribavirin, Remdesivir, lopinavir/ritonavir, antibiotics such as azithromycin and doxycycline. COVID-19 patients also had received anti-parasitic drugs such as ivermectin.

DEXAMETHASONE FOR COVID-19 PATIENTS

A collaborative study on dexamethasone was led by Dr. Horby *et al.* and published in the online edition of NEJM on July 17, 2020. The study was funded by the UK National Health Service (NHS).

RESULTS

A total of 2104 patients received dexamethasone, and 4321 received the usual care.

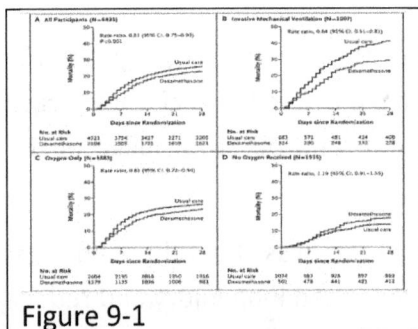

Figure 9-1

The dexamethasone group had 482 deaths (22.9%) compared with 1110 deaths (25.7%) in the usual care group within 28 days after randomization (age-adjusted rate ratio, 0.83; 95% confidence interval [CI], 0.75 to 0.93; P<0.001).

In the ventilator-dependent patients, the dexamethasone group death rate was 29.3% compared to the control group, which was 41.4%. (rate ratio, 0.64; 95% CI, 0.51 to 0.81). (Figure 9-1)

Among those on oxygen without mechanical ventilation, the mortality in the dexamethasone group was 23.3% versus 26.2% in the control group (rate ratio, 0.82; 95% CI, 0.72 to 0.94).

Those patients who did not need oxygen or mechanical ventilation did not benefit from the dexamethasone treatment. The death rate in the dexamethasone group was 17.8% v. 14.0% in the control group (rate ratio, 1.19; 95% CI, 0.91 to 1.55).

SECONDARY OUTCOMES.

The risk of progression to invasive mechanical ventilation was lower in the dexamethasone group than in the usual care group (risk ratio, 0.77; 95% CI, 0.62 to 0.95) (Figure 9-2)

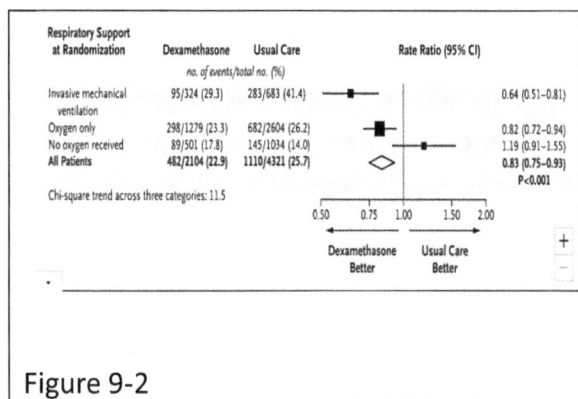

Figure 9-2

Table 2. Primary and Secondary Outcomes.

Outcome	Dexamethasone (N=2104)	Usual Care (N=4321)	Rate or Risk Ratio (95% CI)*
	no./total no. of patients (%)		
Primary outcome			
Mortality at 28 days	482/2104 (22.9)	1110/4321 (25.7)	0.83 (0.75–0.93)
Secondary outcomes			
Discharged from hospital within 28 days	1413/2104 (67.2)	2745/4321 (63.5)	1.10 (1.03–1.17)
Invasive mechanical ventilation or death†	456/1780 (25.6)	994/3638 (27.3)	0.92 (0.84–1.01)
Invasive mechanical ventilation	102/1780 (5.7)	285/3638 (7.8)	0.77 (0.62–0.95)
Death	387/1780 (21.7)	827/3638 (22.7)	0.93 (0.84–1.03)

* Rate ratios have been adjusted for age with respect to the outcomes of 28-day mortality and hospital discharge. Risk ratios have been adjusted for age with respect to the outcome of receipt of invasive mechanical ventilation or death and its subcomponents.
† Excluded from this category are patients who were receiving invasive mechanical ventilation at randomization.

Figure 9-3

CONCLUSIONS

The authors concluded dexamethasone was useful in reducing mortality in COVID-19 patients who were receiving oxygen and or were on mechanical ventilation. (Figure 9-3)

WHAT IS HYDROXYCHLOROQUINE?

Hydroxychloroquine, an antimalarial and immunomodulatory agent, has shown antiviral activity against SARS-CoV-2. It exerts antiviral activity by increasing intracellular pH resulting in decreased phagolysosome fusion, impairing viral receptor glycosylation. It also decreases the production of cytokines, especially IL-1 and IL-6, by inhibiting toll-like receptor signals. It might also have a potential anti-thrombotic effect.

A small open-label trial from France reported a faster rate of SARS-CoV-2 clearance after six days of hydroxychloroquine alone or hydroxychloroquine in combination with azithromycin treatment.

Several prior hydroxychloroquine studies reported conflicting results, which forced the FDA to revoke the prior emergency use authorization (EUA) in July 2020.

As of July, there were several NIH randomized trials of HCQ underway. Based on the early reports, the authors included HCQ alone or in combination with azithromycin in treating COVID-19 patients.

HYDROXYCHLOROQUINE – HENRY FORD STUDY

Samia Arshad *et al.* tested the role of hydroxychloroquine therapy alone and in combination with azithromycin in hospitalized patients positive for COVID-19.

STUDY GROUP

They included all the COVID-19 related admissions between March 10, to May 2, 2020. They studied the in-hospital mortality among those who received hydroxychloroquine (HCQ) alone, hydroxychloroquine in combination with azithromycin, azithromycin alone, or neither. There were

2541 patients with a mean hospitalization of six days. The patients received HCQ: They recommended 400 on HCQ mg twice daily on day 1, and 200 mg twice daily for 2-5 days. They monitored QTc intervals on cardiac patients and those with QTC >500 ms. Other treatments included corticosteroids and Tocilizumab.

OUTCOMES

Overall mortality was 18.1% in the entire study group, 13.5% in the HCQ-alone group, 20.1% among those receiving HCQ + azithromycin, 22.4% among the azithromycin alone group, and 26.4% for neither drug (p < 0.001). Overall, in-hospital mortality of 18.1% reflected a high prevalence of co-morbid conditions in COVID-19 patients admitted to their institution. (Figure 9-4).

Drugs	Mortality	CI
Overall	18.1%	(95% CI: 16.6%–19.7%)
HCQ + Azithromycin	20.1%	(95% CI: 17.3%–23.0%)
HCQ alone	13.5%	(95% CI: 11.6%–15.5%)
Azithromycin alone	22.4%	(95% CI: 16.0%–30.1%)
Neither drug (control)	26.4%	(95% CI: 22.2%–31.0%)

Figure 9-4

In the multivariable Cox regression model of mortality using the control group as the reference, the hydroxychloroquine alone group had a 66% (p < 0.001) decreased mortality hazard ratio, and hydroxychloroquine + azithromycin decreased the mortality hazard ratio by 71% (p < 0.001). The primary cause of mortality was respiratory failure (88%). None had documented Torsades de pointes.

The predictors of mortality were age >65 years (HR:2.6 [95% CI: 1.9–3.3]), white race (HR:1.7 [95% CI:1.4–2.1]), chronic kidney disease (HR:1.7 [95% CI:1.4–2.1]), reduced O2 saturation level on admission (HR:1.5 [95% CI:1.1–2.1]), and ventilator use during admission (HR: 2.2 [95% CI:1.4–3.3]). Most patients (52%, n = 1,250) had BMI ≥ 30.

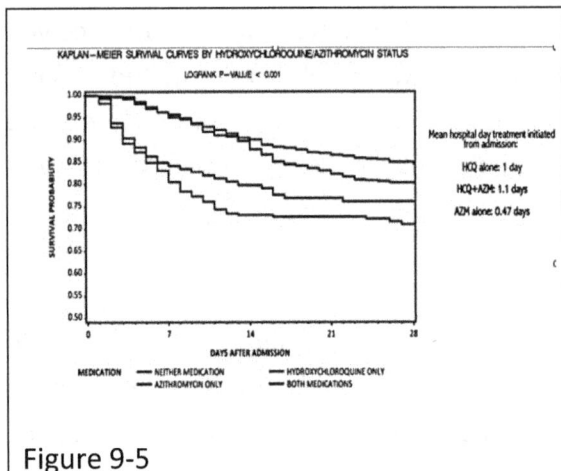

Figure 9-5

Kaplan–Meier survival curves showed significantly improved survival among patients in the hydroxychloroquine alone and hydroxychloroquine + azithromycin group compared with groups not receiving hydroxychloroquine and those receiving azithromycin alone. The survival curves suggest the enhanced survival in the hydroxychloroquine alone group persisted up to 28 days from admission. (Figure 9-5)

CONCLUSIONS

Hydroxychloroquine alone and in combination with azithromycin reduced COVID-19 associated mortality.

They related the benefit of HCQ to its use early in the disease course with standardized doses.

The benefits were less likely when used late in the disease course when patients had already experienced hyperimmune response or critical organ damage.

These findings agreed with the recent NIH guidelines showing a potential role for hydroxychloroquine in the treatment of hospitalized COVID-19 patients without co-administration of azithromycin.

COUNTERPOINTS

This was a retrospective, non-randomized, non-blinded study design. We need prospective trials to examine that impact.

A higher percentage of people received steroids in the treatment group than in the control group. If they showed an improvement in mortality in the treatment arm using a combination of HCQ and steroids in the initial stages, what is wrong with that? Anything that safely reduced the mortality was significant, as we did not have any treatment that had shown such a dramatic improvement in outcomes.

No adverse cardiac events like significant ventricular irregular beats or serious arrhythmias were reported in other studies.

The HCQ group had no Torsades de pointes. The researchers postulated that timely treatment with HCQ prevented the development of myocarditis. Close patient monitoring and established electrolyte protocols also reduced the chances of serious dysrhythmias.

In this study, 82% received HCQ within the first 24 hours of admission, and 91% within 48 hours of admission.

The mortality rates among those requiring mechanical ventilation ranged from 30% to 58%

HENRY FORD HOSPITAL HCQ STUDY

METHODS: The authors conducted an open-label controlled trial in patients hospitalized with COVID-19. They randomly assigned patients to the treatment group, and patients received oral or intravenous dexamethasone (six mg once daily) for up to ten days, while the control group received standard care. The primary endpoint was 28-day mortality.

Enrollment, Randomization, and Inclusion in the Primary Analysis.

The study period was between March 19 and June 8, 2020. Of 11,303 screened patients, 6425 were enrolled in the study (2104 in the dexamethasone and 4321 in the control group). Azithromycin use in the dexamethasone group and the usual care group (24% vs. 25%) was similar. Hydroxychloroquine use ranged from 0 to 3%. Only five patients were given remdesivir.

LIMITATIONS OF MOST STUDIES AND NUMEROUS META-ANALYSES OF THOSE STUDIES.

- Most were retrospective studies
- Limited randomized placebo-controlled trails
- Small study groups
- Vastly varying drug dosages among the studies
- No standardized time duration or endpoints among the studies
- Many patients received the drug in the advanced stages of the disease when many of them were in critical conditions
- The statistical analyses did not include or account for many confounding factors
- None of the studies included the patients at the time of the diagnosis

Hence, reviewing ten poorly designed, ill-conceived, and vastly different studies and performing a meta-analysis does not improve the reliability of the data. Keeping these points in mind, let us review these three meta-analyses

ENDPOINT DEATH RATES META-ANALYSIS

A meta-analysis of 29 studies of hydroxychloroquine (HCQ) with or without Azithromycin (AZM) was reported on August 26 in Clinical Microbiology and Infection. Only three studies were randomized studies. They dropped twelve studies for several reasons. Even though HCQ reduced mortality by 17%, it did not reach statistical significance due to the sample size. The HCQ and AZM group showed a 27% *increase* in mortality. It was concluded HCQ with or without AZM did not provide mortality benefit. However, as the 29 studies were poorly designed, a combined analysis cannot prove or discredit anything.

PROGRESSION META-ANALYSIS

A meta-analysis of six placebo-controlled studies on disease progression was published on August 18 in the Journal of Virology. There was no improvement in the treatment groups in terms of viral clearance, clinical symptom progression, gastrointestinal side effects, or mortality. There were no differences in the skin side effects and cardiac side effects.

HCQ SAFETY PROFILE META-ANALYSIS

A meta-analysis of HCQ in COVID-19 looking at safety profile was published on August 11 in the European Journal of Clinical Pharmacology. Ten studies used more than 4000 mg of HCQ. They concluded that HCQ increased gastrointestinal symptoms. One VA study showed that cardiac arrhythmias increased in the HCQ group.

There were hundreds of small- and medium-sized studies that were all poorly designed and had inconclusive results.

However, epidemiological evidence points to a different inference. Take, for example, India, where HCQ is readily available to most people at a very nominal cost. Many people were using 200 mg HCQ as prophylaxis twice a week. It might have been one reason, along with many others, for a much-reduced mortality rate in India despite many positive cases.

Many medical professionals I have communicated with have attested they benefited from HCQ in the early phases of the COVID-19 infection.

Elavarasi *et al.* reported in MedRXiv on a meta-analysis of twelve observational and three randomized trials between December 2019 and June 8, 2020. The study involved 10,659 patients, of whom 5713 received Hydroxychloroquine (HCQ). They found no significant reduction in mortality (RR 0.98), time to fever resolution (Difference—0.54 days), or development of ARDS with HCQ treatment (RR 0.90). There was a higher incidence of EKG abnormalities in the HCQ group (RR 1.46). They found no benefit in using HCQ in the management of COVID-19 patients.

ANTIBODY TRAIL

The Eli Lilly drug company tested three different doses of anti-SARS-CoV-2 antibody LY-CoV555 compared with a placebo in a trial enrolling 450 patients. The middle dose of 2,800 mg met the trial's target of significantly reducing the presence of SARS-CoV-2 after eleven days. Other doses of the antibody-drug, including the 700 mg dose and the 7,000 mg dose, did not meet the goal.

Ref:

https://www.henryford.com/news/2020/07/hydro-treatment-study

https://www.ncbi.nlm.nih.gov/pmc/articles/PMC7330574/

https://academic.oup.com/aje/article/doi/10.1093/aje/kwaa093/5847586

https://www.nejm.org/doi/full/10.1056/NEJMoa2021436?fbclid=IwAR226Auyb5xyoscwKtw
YhRu795Kr-9UfORMZVeq5XX96E9JK7OElRO_KOq0\\

https://www.nytimes.com/2020/08/19/us/politics/blood-plasma-covid-19.html?

https://medium.com/microbial-instincts/the-verdict-of-3-new-meta-analyses-on-
hydroxychloroquine-for-covid-19-40cfa9381e9a

https://www.cnbc.com/2020/09/16/coronavirus-eli-lilly-reports-a-reduced-rate-of-
hospitalization-for-patients-using-its-antibody-treatment.html

https://www.thelancet.com/journals/lanrhe/article/PIIS2665-9913(20)30305-2/fulltext

https://www.medrxiv.org/content/10.1101/2020.07.04.20146381v1

10 ROLE OF ZINC IN COVID-19 PATIENTS

COVID-19 infection had no known treatment. There was a lot of interest in exploring various antiviral agents and immunomodulators to prevent COVID-19 replication and minimize tissue damage. Zinc was one such element that had antiviral properties. Zinc had been tried in many viral infections such as hepatitis C, HIV, and others. Its role in the management of COVID-19 patients was studied.

Zinc was noted in intracellular compartments such as the nucleus (30%–40%), cytosol, and other organelles and specialized vesicles (50%), and the rest was bound with cell membrane proteins. While the cells needed a constant supply of zinc, free zinc ions ($Zn2+$) could be toxic to the cells by inhibiting cytoplasmic enzymes such as adenylate cyclase. The plasma zinc level was 10 to 18 mol/L representing 0.1% of total body zinc. (Figure 10-1)

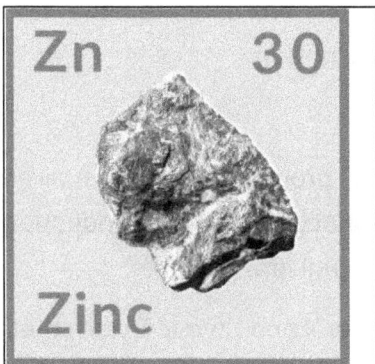

Figure 10-1

Zinc was one of the major factors that controlled the function and proliferation of neutrophils, NK cells, macrophages, T and B lymphocytes, and cytokine production by the immune cells.

Interferons (IFN) and cytotoxic T lymphocytes were essential in clearing the viruses. Zinc ions (Zn2+) took part in the normal development, differentiation, and function of immune cells.

MECHANISM OF ACTIONS

Zinc is an element with a variety of direct and indirect antiviral properties, working through many channels. It has the potential to enhance both innate and humoral antiviral immunity and restore immune cell function in immunocompromised or older patients.

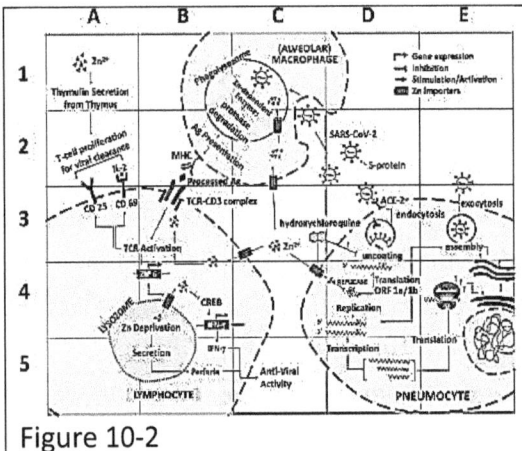

Figure 10-2

It also works synergistically with other antiviral agents and might affix to the virus and unwrap it. Zinc might also stabilize the cell membrane, preventing the virus from getting inside the cell. It also blocks viral multiplication. All these properties suggested a potential benefit of zinc for prophylaxis and treatment of COVID-19 patients. (Figure 10-2)

In COVID-19 patients, zinc worked through both innate and humoral immune responses, stabilized the cell membranes, and inhibited the viral replication by interfering with viral genome transcription.

ZINC DEFICIENCY

Zinc deficiency leads to reduced antibody production and impaired innate immune responses. It also decreases monocyte cytokine production and the chemotaxis and oxidative burst of neutrophil granulocytes.

Zinc deficiency reduces the number of peripheral and thymic T cells, their proliferation in response to phytohemagglutinin, and the functions of T helpers and cytotoxic T cells. It also suppresses the secretion of pro-inflammatory cytokines and thus suppressing inflammatory reaction.

Zinc might also protect or stabilize the cell membrane, which could contribute to the inhibition of the entry of the virus into the cell.

Zinc deficiency leads to increased susceptibility to bacterial, viral, and fungal infections. Patients with a vegan or vegetarian diet, cirrhosis, inflammatory bowel disease, and aging might have low levels of zinc.

Zinc is an effective systemic treatment for viral warts.

CLINICAL STUDIES

Covid-19 primarily infects alveolar cells known as pneumocytes and alveolar macrophages. There are two types of alveolar cells (type I and II). Type I cells provide 95% of the surface area of each alveolus. They are flat; hence, they are referred to as squamous epithelial cells. Type II cells cluster in the corners of the alveoli and have a cuboidal shape.

Infection with SARS-CoV-2 causes an inflammatory condition, also known as pneumonia, affecting alveoli.

Zinc given with chloroquine was more effective. Inside the alveolar cell, zinc reduced the multiplication of the virus.

A retrospective, single-center study, involving 99 COVID-19 patients treated in Wuhan Jinyintan Hospital showed increased concentrations of C-reactive protein, IL-6, and serum ferritin, along with an increased erythrocyte sedimentation rate with zinc treatment. They also showed an increased number of neutrophils with fewer lymphocytes. Cytokine storm patients had similar changes with an overproduction of IL-7, IL-10, GCSF, IP10, MCP1, MIP1A, and TNF-α.

ZINC DOSAGE

Zinc is available as zinc salt, zinc-gluconate, zinc-acetate, zinc-sulfate, and zinc-picolinate. The quantity of elemental zinc in each salt varies. For example, zinc-sulfate contains about 23% elemental zinc. Therefore, to have 50 mg of zinc, you must consume 220 mg of zinc-sulfate.

For adults, the recommended daily allowance is typically 15–30 mg of elemental zinc, though this number varies with age, sex, and health conditions.

Long-term high-dose zinc consumption might cause a decrease in high-density lipoprotein cholesterol levels (HDL), anemia, copper deficiency, and possible genitourinary complications.

HCQ+ AZM+ZINC V. HQC+AZM

Carlucci *et al.* performed a retrospective observational study from March 2 through April 5, 2020, comparing hospital outcomes among patients who received hydroxychloroquine and azithromycin plus zinc (411 patients).

In a univariate analysis, zinc sulfate increased the frequency of patients being discharged and decreased the need for ventilation, admission to the ICU, and mortality or transfer to hospice for patients who were never admitted to the ICU.

After adjusting for the time at which they added zinc sulfate to the protocol, they noted an increased discharge (OR 1.53, 95% CI 1.12-2.09), reduction in mortality, or transfer to hospice care (OR 0.449, 95% CI 0.271-0.744).

However, zinc sulfate did not improve the length of hospitalization, duration of ventilation, or ICU length of stay.

Several other studies found a combination of zinc and hydroxychloroquine given within the first 48 hours to have a favorable outcome.

The Henry Ford study, which was a retrospective study involving more than 2400 patients, showed a favorable result among those patients who received zinc, hydroxychloroquine, and steroids. This is covered in a subsequent chapter.

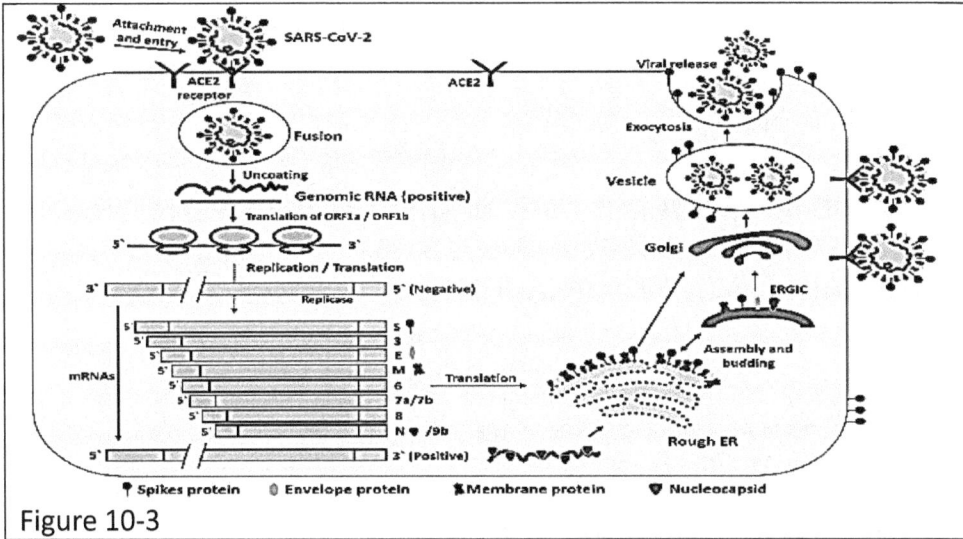

Figure 10-3

Figure 10-4 Foods rich in Zinc

Ref:

https://www.ncbi.nlm.nih.gov/pmc/articles/PMC7247509/

https://www.ncbi.nlm.nih.gov/pmc/articles/PMC7250542/

https://www.prnewswire.com/news-releases/newly-published-outpatient-study-finds-that-early-use-of-zinc-hydroxychloroquine-and-azithromycin-is-associated-with-less-hospitalizations-and-death-301094237.html

https://www.medrxiv.org/content/10.1101/2020.05.02.20080036v1.full.pdf

Nik Nikam
1d · Reopening

COVID-19 PANDEMIC - SHOULD SCHOOLS OPEN FOR IN-PERSON CLASSES IN FALL 2020?
Nik Nikam, MD, MHA, HOUSTON, TX

At a time, when the southern United States regions like Florida, California, and Texas are having a record number of new cases some surpassing that of New York, where the Covid-19 patients are flooding the hospitals, and the national conventions going online; at the other end of the spectrum, there is a spirited debate on whether the schools for K1-12 grades should open f...
See More

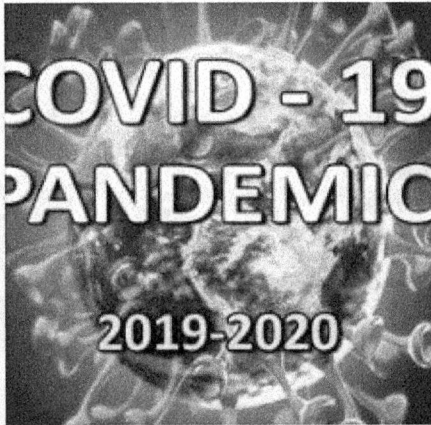

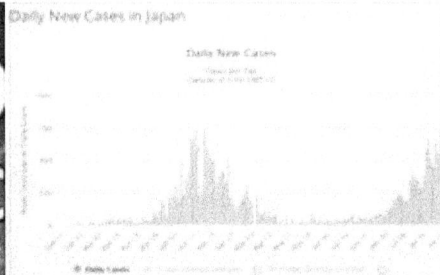

61   210 Comments

11 PLASMA TRANSFUSION

The concept of administering plasma is to provide the antibodies in the earliest stages of infection to prevent viral replication, reduce lung damage, minimize symptoms, and decrease morbidity and mortality.

Physicians have used convalescent plasma in the treatment of viral infections during pandemics dating back over 100 years. They based the concept of providing antibodies that reduced viral multiplication, minimized tissue damage, and saved lives. (Figure 11-1)

Doctors used convalescent plasma during the 1918 Spanish 'Flu, where it improved survival rates. However, the results were less promising during the SARS and Ebola outbreaks.

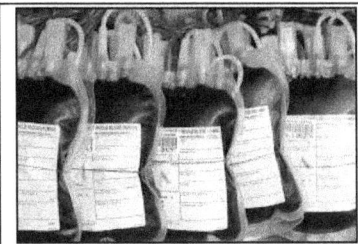

Figure 11-1

ANTIBODIES ARE THE KEY ELEMENTS IN CONVALESCENT PLASMA

Plasma provided passive immunity with neutralizing immunoglobulin IgG, IgM, and IgA antibodies (and possibly other immune mediators) directed against the COVID-19 virus. Antibodies had the potential to prevent a COVID-19 illness, shorten the duration, severity, or prevent its serious or life-threatening complications.

Neutralizing antibodies and spike-protein-binding antibodies appeared within 10 to 15 days of infection. Symptomatic patients had slightly higher levels of IgG than asymptomatic individuals.

OBTAINING PLASMA DONATIONS IN A PANDEMIC

Just as you routinely ask for an autopsy, you develop the habit of looking for potential donors from your hospital patients who recover. Ask if they can donate blood or plasma before they leave the hospital, or later when the need arises. Hospitals were able to use their own patients' blood as they already had their tests, blood types, and other medical data.

I saw calls for plasma donations from family members all over social media when one of their loved ones was seriously ill.

You should use the same guidelines used for blood transfusion, which would protect people's privacy.

On a national level, you want the Red Cross to perform a plasma drive, collect suitable plasma samples, and make them available to all the institutions that have met the criteria to use the plasma in COVID-19 patients.

WHO CAN DONATE: Individuals can donate convalescent plasma if they meet the criteria for a donation and have fully recovered from COVID-19 for at least two weeks? (See "Blood donor screening: Medical history")

Family and friends of a patient who had recovered from COVID-19 can donate to a general convalescent plasma pool, but they cannot designate it to a specific individual. This policy was in place before the COVID-19 crisis began. Direct donations are discouraged for risk of infection, social pressure by family and friends, and the complexity of coordination.

One study showed immune response and antibody titers were higher in males than in females, in older individuals than in younger individuals, and in hospitalized patients versus non-hospitalized patients.

COVID-19 PANDEMIC 2019-2020

They required Individuals wishing to donate convalescent plasma to disclose their vaccination history in the past year to determine eligibility.

Approaches to donor recruitment vary depending on the region of the world, type of health care system, and local disease prevalence.

Major strategies used for COVID-19 convalescent plasma include:

- Donor self-identification based on public awareness about a potential role for convalescent plasma
- Referral of patients who tested positive for COVID-19 infection
- Social media promotions
- Formal news outlets

Patient confidentiality and privacy should be the central focus when recruiting individuals for plasma donations.

Novel approaches for procuring an adequate supply of antibodies for future outbreaks or a second wave might include:

- Banking plasma for future use
- Immunizing selected individuals to generate high-titer units
- Generating immune globulin from animals exposed to the virus
- Using lymphocytes in cultures to produce monoclonal or polyclonal antibodies

Blood banks strongly preferred apheresis for collection for several reasons:

- A single apheresis donation can provide two to four units of convalescent plasma, versus one unit of plasma for one unit of donated whole blood
- People can undergo apheresis as often as twice in seven days, following the US Foods and Drug Administration guidance
- Donor safety: Since apheresis only removes plasma and not red blood cells, it does not make an individual anemic

87

ANTIBODY TITERS

The optimal titer threshold above which plasma is likely to be effective was challenging to determine and was unknown for COVID-19. Concentrating plasma would not provide a higher titer.

Testing the donor during a pre-donation visit was more practical. They could eliminate those who did not have a sufficient titer. It is unknown how much time can elapse between donor testing and the reduction of their titer. It was reasonable to reassess titer at least every two weeks.

Convalescent plasma donors were subjected to the same eligibility criteria used for all blood donors for infectious disease screening. They did not test the plasma for severe acute respiratory syndrome COVID-19 (SARS-CoV-2), the virus that caused COVID-19, as plasma routinely did not transmit these viruses.

Pathogen inactivation methods included solvent/detergent treatment or photochemical inactivation, the latter involving the addition of lipid- or nucleic acid-binding compounds activated by ultraviolet or visible light. The US blood centers did not use this method.

Blood type: They routinely determined ABO and Rh blood types, and plasma transfusion was determined on ABO-compatibility. Some institutions allowed for plasma transfusion in selected cases from a different blood group.

Convalescent plasma carried transfusion reaction risks, like those of standard plasma. Even though antibody-dependent enhancement (ADE) was rare, it was still a theoretical risk. Antibody-containing products might interfere with vaccine efficacy, but this question might arise when the vaccines become readily available.

The collection of the convalescent plasma by plasmapheresis takes one to two hours, using two peripheral intravenous lines. The donor's body replenished the plasma proteins within two weeks.

CHARACTERISTICS OF GOOD CONVALESCENT PLASMA INCLUDED:

•Sufficient titer of the relevant antibody (See 'Antibody titer' below)

•Lack of infectious particles (See 'Infectious disease screening and pathogen inactivation' below)

•Demonstrated safety and efficacy

Convalescent plasma can be transfused to a recipient immediately after collection, as one or two units, or frozen for later administration

CONVALESCENT PLASMA IN THE CLINIC

The first randomized trial of convalescent plasma, reported in China, enrolled 103 patients with severe COVID-19 infection. The median interval between the onset of disease and plasma administration was 30 days. The original target was to recruit 200 patients; however, the study ended early as they could not recruit enough patients. Those who received convalescent plasma had trends towards better outcomes compared with controls; however, this trend did not reach statistical significance.

Mayo Clinic reported the successful transfusion of over 70,000 units of convalescent plasma in COVID-19 patients. Convalescent plasma was more effective when given early in the disease course before patients developed lung and vascular complications.

There were many small-scale observational studies that showed a mortality benefit in the treatment groups compared with the control groups. However, observational studies are constrained by the small size, and various sources of bias related to patient selection, co-interventions, data analysis, studies are not consistent regarding the dose, titer, and timing of plasma administration.

Convalescent plasma could be given in concert with other therapies, including antiviral agents and steroids.

Some institutions used antihistamines and/or acetaminophen as pre-medications before plasma transfusions.

Most clinical studies reported giving one to two units of plasma (250 mL per unit). The antibodies might remain effective for several weeks to a few months. However, multiple infusions may increase the risk of transfusion-associated circulatory overload (TACO), particularly in individuals with complicated pulmonary disease. (See "Transfusion-associated circulatory overload (TACO) or transfusion-related acute lung injury (TRALI)").

The required frequency of repeat dosing with convalescent plasma as prophylaxis for high-risk individuals (rather than as treatment) was unclear.

EXPANDED ACCESS PROGRAM

An expanded access program (EAP) through the US Foods and Drug Administration (FDA) was the major mechanism for obtaining convalescent plasma in the US.

All institutions were eligible to take part in the EAP. It provided for the administration of convalescent plasma to COVID-19 patients with a severe or life-threatening disease or who were at risk of progression to severe or life-threatening disease. The Mayo Clinic was the major data coordinating center for the EAP.

Rapid access was also available through an emergency investigational new drug (eIND) pathway for individual institutions and patients with severe or life-threatening COVID-19 patients.

Information about both programs is provided on the FDA website: https://ccpp19.org.

An update from the COVID-19 Convalescent Plasma Project (CCPP) covering 22,000 recipients of convalescent plasma determined the risk of serious adverse events was low. The complications included:

- Transfusion reactions in 89 (<1 percent)

- Thromboembolic complications in 87 (<1 percent)

- Cardiac events in 680 (~3 percent)

Most thromboembolic and cardiac events were unrelated to the plasma.

Some suggested the administration of convalescent plasma earlier in the disease course might have the potential to cause ADE, and that those patients need close monitoring. Monitoring is challenging, as ADE is a clinical phenomenon without specific laboratory findings. Some individuals might experience clinical deterioration unrelated to ADE.

Interference with Vaccinations: Theoretically, administration of plasma containing antibodies against severe acute respiratory syndrome COVID-19 (SARS-CoV-2) could interfere with the ability of the recipient's immune system to recognize and make antibodies against SARS-CoV-2, thus perhaps reducing vaccine efficacy. That was a theoretical concern. Any such data for SARS-CoV-2 was unknown. For other viral infections for which vaccinations are available, the Centers for Disease Control (CDC) noted the antibody-containing blood products might interfere with some but not other vaccines (measles, mumps, and rubella [MMR], but not live-attenuated influenza).

MAYO CLINIC EXPERIENCE

The Mayo Clinic conducted a multicenter, open-label Expanded Access Program (EAP) study including 2807 acute care facilities exploring the efficacy of convalescent plasma in COVID-19 patients. They reported on 35,322 patients who received one or two units of convalescent plasma right after the diagnosis of COVID-19.

The study period was between April 4 and July 4, 2020. They looked at day seven and day 28 mortalities. The patient in the ICU accounted for 52.3% of the cases. They had 27.5% of the patients on mechanical ventilation.

The seven-day mortality rate was 8.7% [95% CI 8.3%-9.2%] in patients transfused within three days of COVID-19 diagnosis, but 11.9% [11.4%-12.2%] in patients transfused four or more days after diagnosis (p<0.001). Similarly, the 30-day mortality was 21.6% vs. 26.7%, with a p<0.0001.

The mortality rates correlated with the Plasma IgG levels:

IgG level	Mortality
>18.45 S/Co	8.9%

4.62 to 18.45 S/Co   11.6%

<4.62 S/Co   13.7%

Based on this data, convalescent plasma significantly reduces mortality in COVID-19 patients. The reduction is proportional to the early administration of plasma with IgG antibodies levels >18.45 S/Co.

INDIAN EXPERIENCE

Anup *et al.* reported on a multicenter, open-label, parallel-arm phase II randomized controlled trial conducted in 39 hospitals across India between April 22 and July 14. The study involved 464 patients, 235 assigned to the convalescent plasma (CP) group, and 229 to the placebo group.

The interventional group had moderately ill patients with an oxygen saturation of <93% in room air. They received 200 mL of CP and conventional treatment, while the control group received conventional treatment alone.

The composite primary endpoint of progression to severe disease (PaO2/FiO2<100) or all-cause mortality at 28 days post-enrollment was noted in 18.7% in the intervention group versus 17.9% in the control group.

There was no benefit in the overall mortality in the treatment group (13.6% v. 14.6%) [OR 1.06 95% CI -0.61 to 1.83]

Counterpoints. It was not clear from the study when they administered the plasma. In the Mayo Clinic study, the patients received plasma within 48-hours after the diagnosis of COVID-19 infection. They administered plasma before the patients developed pulmonary complications or desaturation. They also measured the antibody titers to ensure the levels were adequate.

In the Indian study, patients received plasma at later stages, when their oxygen levels were below 93%. Also, there was no mention of antibody titers, and it was uncertain whether the plasma had adequate antibody levels. The Mayo study showed the survival rates proportional to the time of administration from COVID-19 onset and adequate antibody levels.

COVID-19 PANDEMIC 2019-2020

Ref:

Useful resource on Plasma transfusion for COVID-19 patients.

https://www.fda.gov/emergency-preparedness-and-response/coronavirus-disease-2019-covid-19/donate-covid-19-plasma

https://covidplasma.org/

https://ccpp19.org/

https://www.hematology.org/covid-19/covid-19-and-convalescent-plasma

https://newsnetwork.mayoclinic.org/discussion/calling-all-covid-19-survivors-now-you-could-help-others-defeat-it/

https://www.wsj.com/articles/national-study-finds-convalescent-plasma-to-treat-covid-19-is-safe-11589453869

https://www.medrxiv.org/content/10.1101/2020.08.12.20169359v1

https://www.houstonchronicle.com/news/houston-texas/article/Houston-Methodist-first-in-the-nation-to-try-15164229.php

https://www.medrxiv.org/content/10.1101/2020.09.03.20187252v2

https://www.medrxiv.org/content/10.1101/2020.09.03.20187252v1

Diagnostic Test Sensitivity in the Days After Symptom Onset[†]

SARS-CoV-2 Test	Days after Symptom Onset		
	1—7	8—14	15—39
RNA by RT-PCR	67%	54%	45%
Total Antibody	38%	90%	100%
IgM	29%	73%	94%
IgG	19%	54%	80%
Adapted from: Zhao J et al. Antibody responses to SARS-CoV-2 in patients of novel coronavirus disease 2019. *Clin Infect Dis*. 2020 Mar 28.[29]			

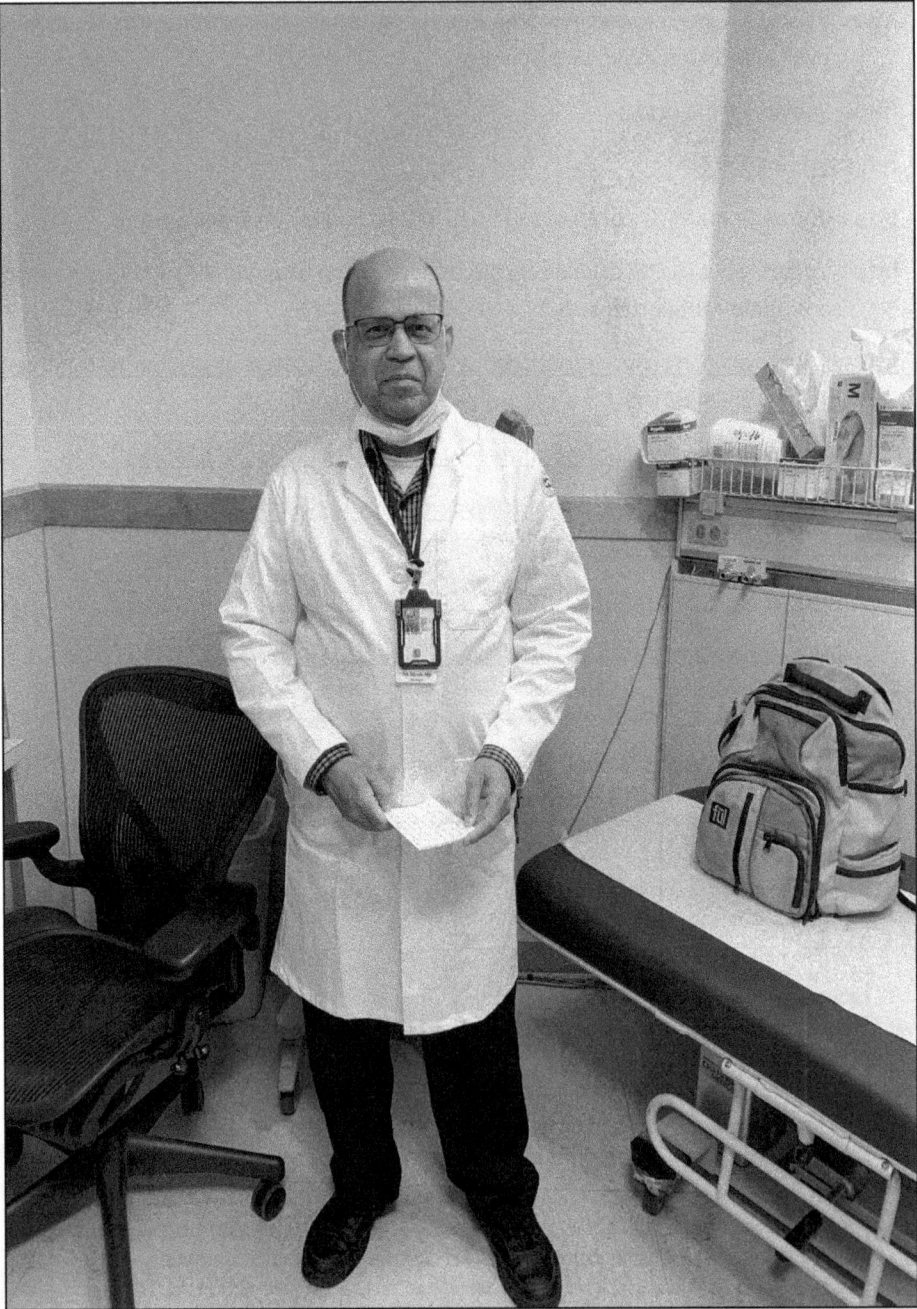

12 DEATH RATES FROM COVID-19

In the beginning, the death rate for COVID-19 was unknown, but the data suggested it could be ten times higher than for the 'flu. It was noted to be higher among vulnerable groups like the elderly, people with hypertension, obesity, heart disease, and people with weakened immune systems.

DOUBLING OF DEATH RATES

Look at the New York times death rates chart based on the countries. It tracked how fast the death rates were doubling among various countries. Spain led in the race by doubling every 2-3 days. France's death rate was doubling close to every three days. In the US, the death rate was doubling every 4-5 days. Italy doubled the death rate every day.

Often people go to the hospital with 'flu-like symptoms and then develop a worsening of existing conditions such as heart and kidney failure. When they die, cardiac arrest due to heart failure is listed as the principal cause of death along with HTN, CAD, CKD, etc. Rarely is the 'flu listed as the primary cause?

In some countries, however, COVID-19 was reported as a cause of death in the presence of other conditions. In Italy, anyone who died and tested positive for COVID-19 was included in the fatality rate, whether COVID-19 was the primary cause of death or not.

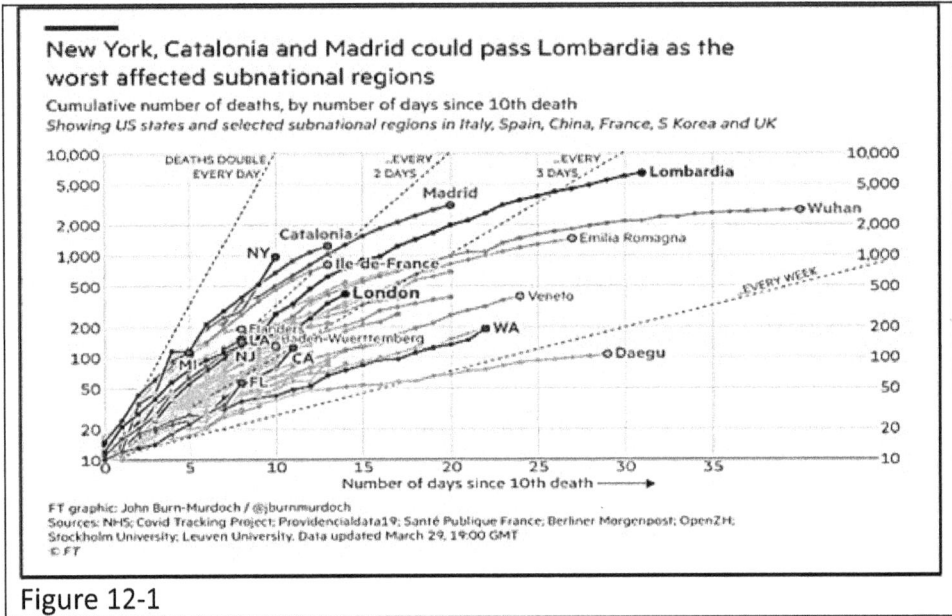

Figure 12-1

In Germany, COVID-19 was not included as the cause of death if patients had comorbid conditions. This highlights the disparities in COVID-19 death rate reporting between countries, as no standard method for reporting was established. In Italy, the fact that the population age is higher on average than other countries may have further inflated the COVID-19 death rate reporting as COVID-19 primarily infected the elderly, and this same population is more likely to have underlying health conditions. (Figure 12-1)

DEATH RATES BASED ON AGE

The figure of death rates based versus demonstrates that 99% of deaths were in patients aged 50 years or older. The highest rate was in people aged 80 years or older. This figure shows what to expect when you are dealing with older patients with multiple medical comorbid conditions like heart failure, hypertension, kidney disease, etc.

COMORBID CONDITIONS AND THEIR EFFECTS

AGE	DEATH RATE confirmed cases	DEATH RATE all cases
80+ years old	21.9%	14.8%
70-79 years old		8.0%
60-69 years old		3.6%
50-59 years old		1.3%
40-49 years old		0.4%
30-39 years old		0.2%
20-29 years old		0.2%
10-19 years old		0.2%
0-9 years old		no fatalities

Figure 12-2

In the US, common comorbid conditions included cardiovascular disease (13.2% mortality), diabetes (7.3% mortality), chronic respiratory diseases (6.3% mortality), hypertension (8.4% mortality), and cancer (7.5% mortality). (Figure 12-2)

The Italian study showed when the patient had three or more conditions, mortality jumped to more than 50%. When people had no pre-existing conditions, mortality was less than 1%. (Figure 12-3)

PRE-EXISTING CONDITION	DEATH RATE confirmed cases	DEATH RATE all cases
Cardiovascular disease	13.2%	10.5%
Diabetes	9.2%	7.3%
Chronic respiratory disease	8.0%	6.3%
Hypertension	8.4%	6.0%
Cancer	7.6%	5.6%
no pre-existing conditions		0.9%

Figure 12-3

Among the deaths rates during the recent pandemics, COVID-19 was 2%, compared to 9.6% with SARS and 34% with MERS. (Figure 12-4)

Virus	Death Rate
Wuhan Novel Coronavirus (2019-nCoV)	2%*
SARS	9.6%
MERS	34%
Swine Flu	0.02%

Figure 12-4

Death rates in COVID-19, SARS, MERS, & FLU

Ref:

www.worldometers.com

https://www.worldometers.info/coronavirus/coronavirus-death-rate/

13 CONTACT TRACING: WHEN, WHERE, AND HOW?

Contact tracing is the key to slow the spread of any infection in an epidemic or a pandemic. Contacts are defined as people who had been in contact with an infected person and, thus, potentially exposed to the virus. It is the most challenging and human resource-intensive task of dealing with a pandemic. It might be easier to design and build a ventilator in 30 days and mass produce them to meet the world's requirements than to develop an effective contact tracing system. Planning for contact tracing should begin long before the epidemic hits. The earlier you start, the better the chance for reducing the spread of the virus.

Contact tracing for COVID-19 became much more essential when countries reopened their economies, to ensure the business places were free of people with infections, and to prevent further viral spread.

However, most countries thought that contact tracing was not necessary until the epidemic became out of control. One reason an epidemic gets out of control is due to the lack of means and resources in identifying active cases, isolating them, and performing contact tracing at the beginning of an epidemic. Hence, infected people continue to spread the infection to many others.

Many aspects and outcomes of contact tracing are important to consider:

- ✓ Caseworkers' role
- ✓ Public education
- ✓ Guidance on where people can access help
- ✓ Follow-up on patient welfare
- ✓ Help with access to food for quarantined individuals
- ✓ Digital monitoring of people with active COVID-19 infection
- ✓ Lessons from South Korea, Germany, and Hong Kong

TRAINING OF CONTACT TRACING CASEWORKERS

A medical professional background was not required to take the Caseworker's training for COVID-19 contact tracing. Dr. Emily Gurley, an epidemiologist from John Hopkins, provided a complimentary 6-hour online class introducing people to the concept of contact tracing and how to follow-up step by step. As of August 2020, more than 130,000 people had completed the training. Her comprehensive interview is available via the CNBC link.

The course covered topics such as viral transmission from person to person, communicating with individuals, and maintaining records and confidentiality. There were many other online training courses offered by various institutions, many of them for free, and some with nominal cost.

Many people who had been displaced from their regular place of employment became contact tracers, providing them with an opportunity to learn about the pandemic, share the information with other people, and engage themselves productively, along with making a few dollars. They called this a "win-win situation" in the US. The public health system, the patients, the contacts, and you benefit from this experience and save lives.

WHO HIRES CONTACT CASEWORKERS?

The state and local health departments hire contact workers to follow up with people who have tested positive for COVID-19. Searching for job ads under the public health service and caseworkers or search Google for caseworker yields several advertisements for jobs for COVID-19 contact tracing. Check your local area as they are always looking for people to perform these important and tedious tasks.

Caseworkers do their jobs remotely online. They receive information from the health department and log their responses into the database link provided to them by their employers. Caseworker salaries range from $25,000 to $60,000. It could be a fantastic way to spend your long and lonely hours and days talking to people and helping them with information and knowledge.

CASEWORKERS TASK BEGINS

Once a caseworker gets the information about an active case, the caseworker contacts the person identifies the agency they work for, and enquires about the individual's welfare. The caseworker checks on the living accommodations and meal arrangements in place. If the person needs accommodation, local authorities can provide a hotel room or another safe place, and caseworkers can also arrange for food delivery if necessary. Caseworkers explain what symptoms the person should watch for and the importance of 14 days of strict quarantine.

Caseworkers provide patients with education, information, and support to help them understand their risk, what they should do to isolate themselves from others, and how to monitor themselves for serious symptoms. They educate them on the possibility they could spread the infection to others even if they do not feel ill.

TRACING THE CONTACTS

Next, the Caseworker enquires about all the persons who might have been in contact with the COVID-19 positive person. The Caseworker then contacts them and explains that they might have been exposed to the virus as one of their contacts has tested positive for the virus. They do not disclose a positive person's name. They instruct when to contact family doctors or visit an emergency center.

Caseworkers encourage people who had contact with someone with COVID-19 infection to prevent the further spread of the disease by:

- ➤ Staying home for 14 days.
- ➤ Maintaining social distance (at least six feet).
- ➤ Monitoring their temperature twice daily.
- ➤ Watching for symptoms of COVID-19 like dry cough, headache, and/or shortness of breath.

As one person with COVID-19 infection often encountered many people, it can be a tedious process contacting all those people and informing them of the situation.

WHO recommended finding caseworkers from the local communities as they would be familiar with the people, their language, and culture. They could make contacts quickly, explain in their language, and get better compliance.

REAL-LIFE LESSONS

HONG KONG

Hong Kong had a limited number of new active cases. Whenever someone came from outside the country, they were automatically quarantined for 14 days. They also received an electronic bracelet that tracked their movements. A caseworker checked on them every 3-4 days. They followed similar steps for newly diagnosed cases of COVID-19 patients. Hence, they could reduce the spread of infection effectively.

CHINA

When China had an outbreak of new cases in Beijing, they quarantined 30 blocks close to the Xinfidi market, where the outbreak occurred. They hired 100,000 caseworkers to reach out to all those who were exposed to the coronavirus.

COVID-19 PANDEMIC 2019-2020

NEW ZEALAND

Two sisters returned to New Zealand from the UK to see a sick family member and were quarantines in a hotel. However, they left quarantine saw the family member before their test results for COVID-19 came back. Later their tests came back as positive.

As officials gathered information on all the persons who encountered them from the airport in the UK to the hotel, etc., they produced a list of 320 people. Now, imagine contacting 320 people on two continents! Think about the challenges, the coordination, the resources, and obstacles they had to deal with. But they did it!

GERMANY

Germany had a national Medical Information System (MIS). They entered the data on every person who was noted to have an active COVID-19 infection in the record. They tracked their data on an app. They also had a team of caseworkers who followed any contacts those patients had. Hence, they had exceptional results with quarantine and contact tracing. Their number of cases in June 2020 reflected that.

THE UNITED STATES CHALLENGE

According to estimates, the US needed 100,000 to 300,000 contact workers nationwide at $3.6 billion. However, the CARES Act had allocated only $631 million for the contact tracing project. They based the caseworkers' needs on the population size and the extent of COVID-19 spread. The CDC had recommended empowering state and local authorities to develop caseworkers' programs for contact tracing.

It might have worked in countries with 50 to 100 cases per day. However, when the US was experiencing 40,000 to 45,000 new cases per day, it was a frightening task to contact all those people and all the people they had contacts with.

Both Google and Apple worked on an app to track people. However, it was challenging in the US because of concerns about privacy and government surveillance of citizens.

The biggest challenge faced in the US was a lack of standardization, such as those seen in Hong Kong, South Korea, and Germany. In the US, each state had its own protocols and procedures with varying success.

DIGITAL CONTACT TRACING

Digital tracing of active cases and contacts was not new to the field of epidemic diseases. With the advent of smartphones, computers, and national databanks, it was easy to track who had tested positive and who they had been in contact with. However, there were significant privacy concerns regarding how this data might be used, and many of the tracing apps relied on unreliable connections to Wi-Fi, Bluetooth, or cellular data.

Yet, there were several advantages to the digital system, which could automatically follow the active cases and instruct contacts to reach out to the caseworkers for further instructions and directions.

A single app that can track the entire nation as they did in Germany is most effective. However, in countries like India, with a population of 1.3 billion people, it is particularly challenging to collect data, store the data, and educate the public on how to use the information.

Another significant problem in the western world, such as the UK and the US, was that strict laws protected private healthcare information (HIPAA in the US,). This placed limitations on what information could and could not be on an app. There were serious concerns about the data ending up in the hand of hackers.

Several people expressed serious distrust with the health information they were receiving from the CDC, WHO, and political leaders. These, coupled with concerns regarding freedom of speech and expression, rendered it impossible to implement a nationwide digital tracking system in the US and expect people to adopt such a model. Apple and Google had worked on producing an app that could help with contact tracing. However, given the climate in the US, it was an arduous task, even if the apps worked.

It was hard to educate people and difficult to convince them, even though the app-tracking of contacts would have significantly reduced the viral spread.

The alternative was to hire caseworkers to do the job at the ground level. They also faced similar challenges when they contacted people who did not want to take part and give their personal information to a stranger on the phone.

The US needed an alternative, complementary approach to offset the challenges they particularly faced regarding contract tracing. Hence, they were left with more questions and no real solutions.

Ref:

cdc.gov

https://www.cnbc.com/2020/06/22/coronavirus-contact-tracing-will-save-lives-if-officials-build-it.html

https://www.cnbc.com/2020/06/07/coronavirus-contact-tracing-has-become-a-fast-growing-job-opportunity.html

https://www.jhsph.edu/faculty/directory/profile/2895/emily-s-gurley

https://www.cnbc.com/video/2020/06/11/watch-healthy-returns-the-path-forward-with-dr-emily-gurley.html

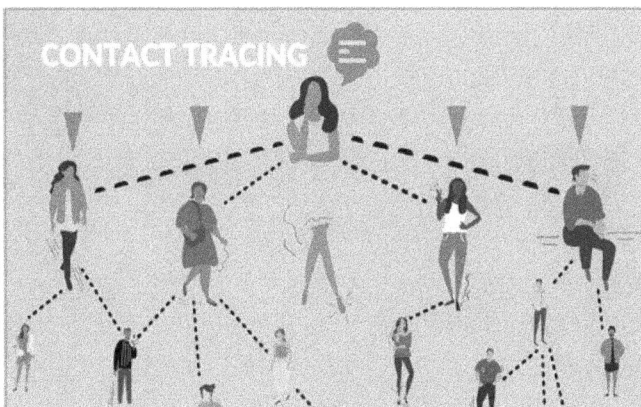

Nik Nikam
March 28 · frontlines update

COVID-19 WORLD NEWS. MARCH 28, 2020

Nik Nikam, MD, MHA, DTM. HOUSTON, TEXAS

WHERE IS THE US IN THIS CORONAVIRUS PANDEMIC ON MARCH 28, 2020
... See More

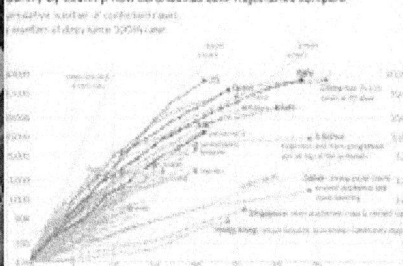

United States

Coronavirus Cases:
104,205

Deaths:
1,701

Recovered:
2,525

Nandana Kanara, Jose Sergio Rizal and 335 others   216 Comments

👍 Like   💬 Comment

14 VACCINE 101 FOR COVID-19

Dr. Jonas Salk developed an attenuated vaccine to combat the rising prevalence of crippling poliomyelitis (polio) around the world. He announced to the public his results from the preliminary trial in 1955. The FDA licensed his vaccine within two hours. And you know the rest of the story. Polio, one of the most debilitating diseases, was eliminated from the world.

Between 1918-1919, the Spanish 'Flu killed more than 50 million people. If they had a vaccine in time, the history of the pandemic would be different.

Vaccines have saved hundreds of millions of lives over the decades. Vaccines provide immunity to the host against a virus or any other organism, preventing a full-blown infection. Vaccines carry a minimal risk for side effects that could be permanent. However, from a public health perspective, vaccines effectively and safely prevent infectious diseases from spreading to millions of people. Vaccination eliminates or minimizes symptoms and reduces the risk of any lethal complications from the infection.

A vaccine stimulates the human immune system to make antibodies against a specific disease or virus, usually with a dead or weakened form of the germ. Later, if you encounter the same germ, your immune system knows what to do. The vaccine provides immunity, so you do not get sick from the same germ. Even if you get ill, the symptoms and the disease course is much milder than it otherwise would be.

When you encounter a new virus, there is usually no immune response or known treatment. The best approach to COVID-19 was to practice prevention by reducing contact with infected persons. Scientists studied the structure and nature of the virus and tried to develop a vaccine that neutralizes its ability to multiply and infect other persons; thus, preventing the spread of deadly infection throughout society, causing millions of causalities.

Vaccine development is complex and very time-consuming. The researchers must make sure a vaccine is effective, safe, and able to be manufactured and distributed to millions of people.

The 'flu vaccination needs new research every year to include the new strains of the 'flu virus, testing, and manufacturing to be ready before the next 'flu season.

A new vaccine is developed chemically or in the laboratory using a virus or one of its key components. Once the substance is extracted, it is developed, tested in laboratory animals for its efficiency to block viral reproduction and its safety (pre-clinical study). Preclinical tests determine the approximate dose ranges and proper drug formulations before proceeding to the Phase 1 trial.

China shared the genetic sequence of the COVID-19 on Jan 10, and scientists have been in a sprint since then to create a viable vaccine. COVID-19 vaccine researchers have ventured into unchartered territories such as using DNA and RNA technologies along with established modalities using an inactivated virus or viral vectors.

Most COVID-19 vaccines tested to date produce minor side effects like fever, chills, and pain at the injection site. Hence, the researchers also used an old tested vaccine such as a meningococcal vaccine as a control, which also creates similar symptoms. Also, researchers identified five unique types of immune responses to the COVID-19 infection, which might account for varied responses to the infection in different populations. It might help to identify whether one vaccine would cover all strains of the virus.

COVID-19 VACCINE

Over 150 vaccines have been tested around the world for COVID-19. Only six to eight made it to the clinical trials. The unique feature of the COVID-19 vaccine race was the variety of vaccine modalities, the organizations developing them, and the geographic distribution of R&D and manufacturing. The lead vaccine manufacturers were from the US, Europe, and China.

While researchers were racing toward the finish line, they were competing against the virus and time, rather than against each other. They readily shared data and collaborated with academic labs and government institutions to overcome common scientific and logistical challenges.

Even competitors like Sanofi and GlaxoSmithKline joined hands to maximize their expertise in the various stages of vaccine manufacturing and distribution. J&J, Sanofi, and Pfizer could make enormous quantities, while smaller players were to partner with others or buy capacity from CMOs.

However, even if clinical data looks good, governments have a challenging time deciding on whether to approve emergency access to unapproved vaccines. They must balance the safety and efficacy concerns versus urgency to protect the frontline healthcare workers with unknown vaccines.

VACCINE TRIAL PHASES 0-V

Phase 0 trials are the first clinical trials done among people. They study how the body reacts to an agent and its effects. They give a small dose to ten to 15 individuals.

The treatment group receives a pre-determined dose of the new vaccine. The control group receives a placebo, or an adjuvant-containing cocktail, or an established vaccine (a vaccine used to protect against a different pathogen) to study the immune response.

Phase 1: The scientists aim to find the best dose of a new product with the fewest side effects. They test the drug in a smaller group of 15 to 30 patients.

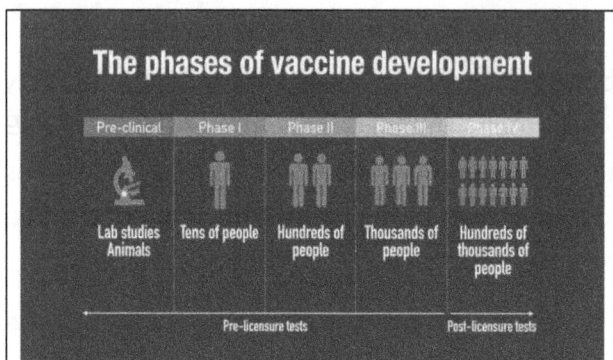

Figure 14-1

They give tiny doses of the drug to a few patients. Then, they try higher doses in other patients until side effects become too severe, or they see the desired effect. The primary aim of the Phase 1 trial is to test a drug's safety. They aim vaccine escalation studies to minimize the chances of serious adverse effects (SAE) by slowly increasing the drug dosage or frequency. (Figure 14-1)

Phase 1 trials can last one to two years and involve fewer than 100 volunteers.

They gather data on antibody production and clinical outcomes (such as illness due to the targeted infection or to another infection). They determine the protective efficacy of the vaccine.

The researchers perform statistical analyzes to establish the significance of the observed differences in the outcomes between the treatment and the control groups. Next, they move to Phase 2 clinical trials.

Phase 2 trials involve larger groups of patients. They could involve several hundred patients. They closely watch patients for side effects and drug efficacy. They test vaccines for their antibody response in a variety of patients, along with their side effects. Once they determine the vaccine is safe and effective, they move to a Phase 3 clinical trial. Phase 2 also includes more vulnerable people at increased risk of contracting the virus or having a serious complication when they get the infection (Figure 14-2). Phase 2 takes at least two years and includes several hundred volunteers.

Phase 3: These trials compare a new drug to a standard-of-care drug. These studies assess the side effects of each drug and which drug works better. Phase 3 lasts three or four years and involves thousands of individuals. Overall, the clinical trial process might stretch to 15 years or more. One-third of vaccines make it from Phase 1 to final approval.

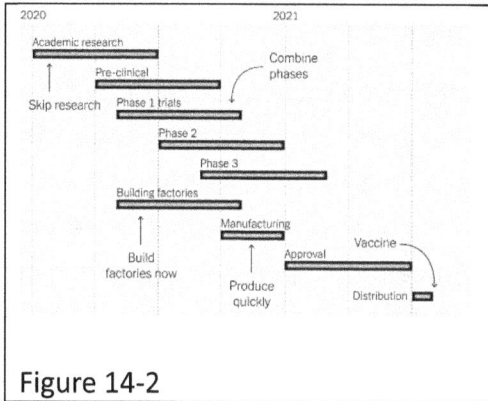

Figure 14-2

They watch patients closely in a Phase 3 study and stop the study if the side effects of the new drug are too severe or if one group has much better results. The FDA requires Phase 3 clinical trials to approve a new drug or vaccine before it releases them to the public. (Figure 14-2)

There can be more than two treatment groups in Phase 3 trials. The control group gets the standard-of-care treatment. The other groups get a new treatment. Neither you nor your doctor chooses the group, and you do not know which group you are in until the trial is over. In the US, the Food and Drug Administration (FDA) approves the vaccines.

Phase 4: Researchers test new vaccines approved by the FDA. They test the vaccine in thousands of patients. They monitor short- and long-term side effects and safety. Infrequently, rare side effects might become known in a large sample of people. They also learn how well the product works and if it is helpful when combined with other treatments.

Vaccine trials might take months or years to complete. It takes time for the antibodies to develop, and the trial must go through all these phases.

Quality Control: Scientists and government agencies track the drug-making process and who gets the vaccine. Safety and side effects are major concerns.

Building Factories: In the past, manufacturers have had to build new factories to mass-produce the vaccine. During COVID-19, factories that could manufacture millions of doses of vaccine in a few weeks were ready and waiting for the FDA to approve the products.

Manufacturing: The leading vaccine developers have already partnered with mega pharmaceutical companies with the technology to manufacture, package, and deliver the vaccines to the various countries in record time. The FDA inspects the factory and approves drug labels.

Distribution: In this modern era, vaccine distribution is easy, but also subjected to abuse. In the US, the federal government planned to use the military to distribute the vaccine in record time to those institutions where the risk of viral spread was greatest. The first group of individuals to receive the vaccine would be frontline workers, healthcare personnel, and nursing home residents, or other vulnerable people. They named the Federal government project "Operation Warp Speed".

Even if they found a vaccine that worked against the new coronavirus, it could be 12 to 18 months before it was ready for public use. That is a fraction of the usual time.

COVID-19 VACCINE CANDIDATES AND THE RACE

One of the world's largest COVID-19 vaccine trials began on July 27, 2020, with plans to enroll 30,000 volunteers. Moderna, a smaller and a new player in the vaccine research, was relying on NIH support for its clinical development and had received almost half a billion-dollar pledge from BARDA.

Volunteers did not know if they were getting the actual vaccine or a placebo. After two doses, scientists tracked infection rates in both groups as they returned to their daily routines, especially in areas where the virus was spreading unchecked. There was no guarantee the experimental vaccine, developed by the National Institutes of Health and Moderna Inc., would provide protection.

The Moderna Phase 1 vaccine trial induced an immune response in all volunteers with mild symptoms like headache, fever, fatigue, and muscle aches. According to the World Health Organization, the Moderna/NIH vaccine was one of 25 vaccines in clinical trials around the world.

Pfizer also began its Phase 3 clinical trial of a COVID-19 vaccine in the US on July 27. It planned to enroll 30,000 volunteers. The volunteers would receive two doses of the vaccine injections or the placebo approximately 28 days apart. Pfizer received a $2 billion grant from the federal government for its vaccine production.

These trials needed multigenerational and multi-ethnic volunteers, who reflected the diverse population. Recruiting 30,000 volunteers for each vaccine trial was an enormous task. However, more than 150,000 people volunteered to take part in an online recruiting questionnaire in the US.

CanSinor in China completed a Phase 2 trial and began a Phase 3 trial; however, the data were not made public. Others, such as the Wuhan Institute of Biological Products and Beijing Institute of Biological Sciences, also had vaccines in Phase 3 trials. Sinovac, Novarox, and Johnson & Johnson had Phases 1 and 2 trials ongoing at the time of writing. The Chinese Academy of Medical Sciences and Inovio has completed a Phase 1 study.

The lead vaccine developers included Pfizer, Johnson & Johnson, and the Jenner Institute at Oxford University. The Oxford group had joined hands with AstraZeneca and had collaborated with Serum Institute of India, the world's largest vaccine manufacturer.

AstraZeneca, a British pharmaceutical company, working in partnership with the Oxford research group, planned to produce more than 100 million doses by September. They planned to produce the vaccine in four places around the world and made a deal with the Serum Institute of India to produce Oxford University's vaccine in India.

The Serum Institute vaccine was undergoing Phase 2 human trials at the time of writing. If successful, the company plans to produce millions of doses in the next few months. It has the potential to produce two billion doses by mid-2021 for worldwide use. (Figure 14-3)

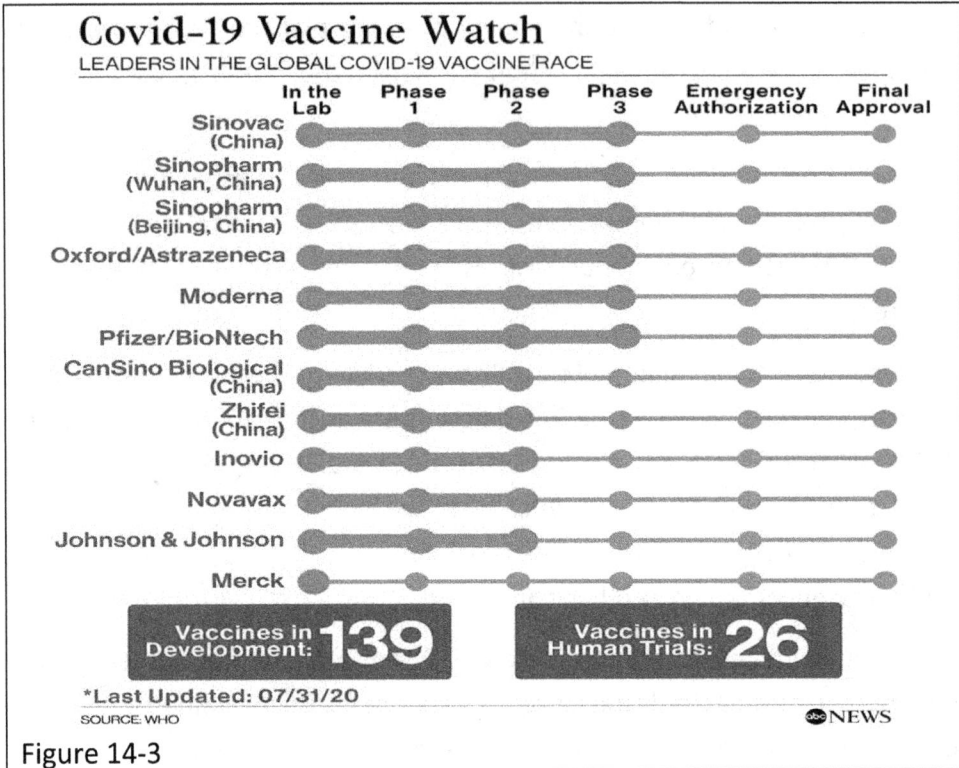

Figure 14-3

On Sept. 1, Canada announced plans to buy 76 million doses of COVID-19 vaccine from Novavax.

THE CUTTER INCIDENT—1955

In 1955, some batches of polio vaccine manufactured by the Cutter company contained the live poliovirus, even though they had passed required safety testing. Over 250 people who received the vaccine developed paralysis. The vaccine was recalled as soon as real polio cases were detected following its use.

The Cutter Incident was a defining moment in vaccine manufacturing history and government oversight of vaccines. It led to the creation of a better regulatory process to ensure vaccine safety. The polio vaccination resumed in the fall of 1955.

ANTIVACCINE GROUP CONCERNS

The polio vaccine saved millions of people from the crippling disease. However, there were people with an opposing view. Occasionally, people might develop unwanted reactions and unacceptable outcomes following vaccination. This, however, can be seen in every branch of medicine—when individuals go for surgery, sometimes they develop complications. Things happen.

You need to look at it from a public health perspective which deals with millions of people, and with COVID-19, several hundred million.

Yes, you might encounter untold reactions to a vaccine. But overall, the public benefit outweighs the individual risk, especially in a pandemic.

Even with the polio vaccine, when the Cutter company had a problem with distributing a live virus, they recognized the problems, fixed the issue, and moved on.

So, you may need some faith. If a vaccine protects millions of people against a lethal disease, you must be prepared to accept a certain risk.

VACCINE WRONGDOING CLAIMS PROCESSING

In the past, there was no system in place to compensate those who had vaccine-related injuries. Now, the National Vaccine Injury Compensation Program (VICP) uses scientific evidence to decide whether a vaccine caused an illness or injury. Compensation is provided to individuals harmed by such a vaccine. The VICP remains a model method for ensuring all persons harmed by vaccines are quickly and equitably compensated while also protecting the companies that make these lifesaving products from dealing with financially unsustainable liability claims through the tort system.

A VACCINE MAY NOT SOLVE ALL CHALLENGES

- Vaccine efficacy needs to be tested and established in larger field trials
- The vaccine might not provide immunity to some individuals and could lead to false hope

- Vaccine manufacturing could hit bumps
- Vaccine distribution could challenge in many countries
- Some people may simply refuse the vaccine

BREAKING NEWS: NOV. 12, 2020

Pfizer announced the Phase-3 trial preliminary results. It found its vaccine against the coronavirus to be 90% effective in reducing the infection rate 7 days after the second dose.

Ref:

https://www.biocentury.com/article/305091/end-of-the-beginning-for-covid-19-vaccines

https://www.biocentury.com/coronavirus.

https://www.nccn.org/patients/resources/clinical_trials/phases.aspx

https://www.cdc.gov/vaccinesafety/concerns/concerns-history.html

https://www.timesnownews.com/health/article/coronavirus-mass-production-by-astrazeneca-begins-100-million-doses-to-be-produced-by-september/602638?

https://www.researchgate.net/publication/341354165_MMR_Vaccine_Appears_to_Confer_Strong_Protection_from_COVID-19_Few_Deaths_from_SARS-CoV-2_in_Highly_Vaccinated_Populations

https://www.webmd.com/lung/covid-19-vaccine#1

https://www.theguardian.com/world/ng-interactive/2020/jul/22/coronavirus-vaccine-tracker-how-close-are-we-to-a-vaccine

https://www.cnn.com/2020/07/27/health/coronavirus-vaccine-trial-begins-moderna-phase-3/index.html

15 HERD IMMUNITY

WHAT IS HERD IMMUNITY?

Herd immunity is a state in a population where most of the people have developed immunity either through infection or through vaccination, thus having protection against any future infection with the same virus.

When a new epidemic or a pandemic begins, nobody is immune; hence, the virus can infect any person who meets a person who has an infection. When everyone is susceptible, the infection can spread to several thousand people a day in a few days or weeks. It was highly unlikely that we could achieve herd immunity in the US, or the world, for the COVID-19 (SARS-CoV-2) in the year 2020.

In the US, we had 6.48 million active COVID-19 cases by Sept. 8, 2020, accounting for 2% of a population of 330 million people.

Even in New York City, which was the epicenter of the COVID-19 pandemic, the immunity rates in September 2020 ranged from just 15% to 21% across various parts of the state despite widespread infection and the highest mortality in the US (33,000 accounting for almost 17% of the US total mortality from COVID-19 infection).

In the first week of April, in Santa Clara, California, only 3% of the people tested had antibodies.

How about getting a controlled infection and getting over it to achieve immunity? That is easier said than done. You do not even wish that on your enemy. COVID-19 is a deadly virus capable of killing even healthy people. It is not possible to know will get severe lung complications that could lead to death. Hence, this method of attaining immunity was not a viable option. We would not have the healthcare resources to deal with an untold surge of new cases flooding the emergency center and hospitals. Also, each infected person would put their entire family at risk—especially elderly people, who are at substantial risk for deadly complications.

The mortality rate among those infected with COVID-19 on The Diamond Princess cruise ship was 1%.

Scientists say that to reach herd immunity, COVID-19 needs to infect between 60% to 80% of the population and that most of those people must develop antibodies. When 80% of the people develop antibodies and immunity, then a person with an active infection could only infect one out of five people as the other four (80%) would have immunity to the virus. The US population is 330 million, and we needed to have 198 to 264 million people exposed to the virus and to have developed the antibodies to the virus to achieve herd immunity. That was not a practical solution, and herd immunity was not essential to mitigate COVID-19 spread.

Increased testing, identifying the active cases, and isolating them and their contacts was a more prudent way to slow the spread of the infection until a vaccine became available. Remember, for every active person you isolate and quarantine, you could prevent the spread of the infection to 500 people over several weeks. That became an essential part of the guidelines when states and countries planned to lift the lockdowns and reopen their economies.

Reopening businesses and schools without a measure to identify active cases and quarantine could spell disaster for the economy if it forced the reintroduction of total lockdown.

After South Korea reopened schools, there was a spike in the number of recent COVID-19 cases, and they had to shut the schools back down. You need to understand that because if you rush to open schools and businesses in a pandemic, the pandemic might be far from over.

Many viral infections like measles, mumps, polio, and chickenpox were common infectious diseases in the past. With vigorous and meticulous vaccination of children, we have been able to achieve herd immunity without the children getting sick with viral infections. We occasionally still see outbreaks in communities that cannot vaccinate or refuse to vaccinate their children.

People receive the 'flu vaccination each year, yet immunity doesn't last long, whereas vaccinations for polio and smallpox provide lifelong immunity. One reason the 'flu vaccination doesn't provide long-term protection is that the 'flu viruses mutate each year, and there are new strains each year that require a modified vaccine.

The presence of antibodies does not mean that people have absolute protection against the same virus. Only time can tell if antibodies to COVID-19 protect people who are exposed a second or third time, whether they develop symptoms or become sick again, or whether the disease course is milder in subsequent encounters. We do not know how long COVID-19 immunity lasts and protects the exposed population.

Keep in mind that false positive and negative studies could alter the statistics. The impact of false positives is bigger when people who are being tested do not have antibodies to the virus, but they test positive. It might provide false hope the person is immune.

To determine true immunity, you need to know how many people have antibodies and how protective these antibodies are in providing immunity.

Based on the number of people tested, over 95% of the American people will not be immune to the new coronavirus. That means most of the US was at risk for COVID-19 as of June 1, even though most of the states had lifted lockdowns and begun to open the economies in stages. Out of ignorance, some people failed to follow the self-mitigation steps laid out by the CDC to wear masks in public places, maintain distance, and frequently wash their hands.

If you do not use the self-mitigation guidelines or lockdown, many people might get infected in a few weeks. This might increase immunity to the virus, but also immensely tax the healthcare system, increase mortality, and create political havoc because of not taking appropriate steps to slow the viral spread and for saving lives.

We saw the impact of the sudden rise in the number of hospital admissions in New York City, which put a tremendous burden on the healthcare system and created the need for thousands of additional ventilators and hospital beds.

The best option in such a situation was to keep the viral spread minimal and hope for a vaccine in 6-18 months. That is how long it was estimated to take to develop a vaccine, test it in animals, and confirm its safety and efficacy in people.

That means even if people return to work, they need to understand over 90% of the population are still at risk for COVID-19 infection and could die from complications. In a pandemic, you might have to learn to accept the self-mitigation steps for 6-12 months or until someone develops a natural immunity through infection or they have a reliable vaccine.

Every time you reopen for the economy, you need to follow:

- ✓ Number of new cases
- ✓ Number of people in the hospital
- ✓ Number of people on the respirators
- ✓ Daily mortality data

SWEDEN WITH NO LOCKDOWN

While the rest of the world was on lockdown and struggling with the massive influx of COVID-19 patients in the hospitals, in the ICUs, and on ventilators, Sweden defiantly remained open. The government issued guidelines for people to stay at home, avoid gatherings of more than 50 people, and closed public places such as museums. Restaurants, bars, parks, and schools remained open. Sweden's ambassador, Karin Olofsdotter, said that they expected to achieve herd immunity by the end of May 2020. The people of Sweden had second thoughts about their government's approach. By late April, they had only ~20% immunity. By the end of May, this number neared 30%.

LESSONS FOR FUTURE PANDEMICS

What were the so-called experts advising during the first two weeks when they had no test kits to detect the people with infections? Nobody talked about distancing, masks, and handwashing, etc. in the beginning. We started doing that *after* we had passed the peak of the pandemic worldwide.

These are excellent lessons and a significant opportunity for Americans to lead the world in the future diagnosis and containment of an epidemic before it becomes a pandemic.

The US should be the lead country in developing testing methods, which were the first checkpoints for viral spread. Developing tests for a new virus helps US citizens and the entire world.

Having diagnostic testing kits would have helped us to detect and quarantine cases during the first 1-2 weeks of the epidemic when 100% of the population was susceptible. That observation should be part of the playbook for the next epidemic or pandemic.

Ref:

https://coronavirus.jhu.edu/from-our-experts/early-herd-immunity-against-COVID-19-a-dangerous-misconception

https://hub.jhu.edu/2020/04/30/herd-immunity-COVID-19-coronavirus/

https://www.jhsph.edu/COVID-19/articles/achieving-herd-immunity-with-covid19.html

https://www.discovermagazine.com/health/is-herd-immunity-our-best-weapon-against-COVID-19

COVID-19 PANDEMIC 2019-2020

https://www.npr.org/2020/05/25/861923548/stockholm-wont-reach-herd-immunity-in-may-sweden-s-chief-epidemiologist-says

16 IMMUNITY PASSPORTS—POST-COVID-19

Let us say you survived COVID-19, got over the dreaded symptoms, and even got a repeat viral test to make sure you were not still spreading the virus. You took an extra step to check the antibody titers. And, you were excited to have a good antibody titer. But, for how long?

The blood banks are accepting your convalescent plasma for other COVID-19 patients. You think you are immune to COVID-19. Then, you question why you should follow the same self-mitigation rules such as six-foot distancing, wearing masks, and frequent hand washings?

Why should the same restrictions apply to you? How would the rest of the world know you have recovered from the coronavirus? How can your coworkers trust your words?

How can recovered people be distinguished from the rest? Should they get an Immunity Passport if they have antibodies and not likely to spread the virus to others? How should they be handled at airports and in other public places?

As of September 9, 2020, more than 27.7 million people worldwide had been diagnosed with COVID-19 with 901, 629 deaths, and 19.8 million recovered. The US reported 6.51 million COVID-19 infections, 194,032 deaths, and 8.8 million recovered. (Figure 16-1)

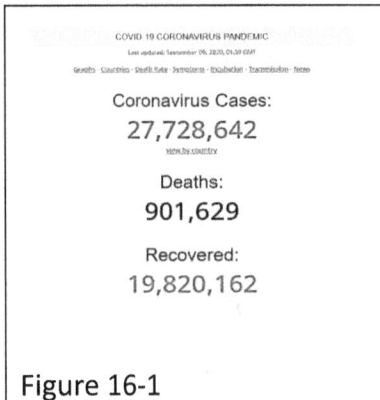

COVID-19 CORONAVIRUS PANDEMIC
Last updated: September 06, 2020, 01:59 GMT
Deaths · Countries · Death Rate · Symptoms · Incubation · Transmission · News

Coronavirus Cases:
27,728,642
view by country

Deaths:
901,629

Recovered:
19,820,162

Figure 16-1

Shouldn't people who had recovered return to full-time duty without lockdown and self-mitigation rules?

We have state driver's licenses and medical licenses with expiration dates that identify individuals. People carry a CPR card that identifies the person and the expiration date.

Why can't people get a national ID card showing they have recovered from COVID-19? Why can't the states issue an ID card with an expiration date to separate the recovered from the infected?

There were many challenges to standardizing the criteria and their reliability. In the future, a new process could be put in place to implement such identification processes.

First, there was no FDA-approved standardization for antibody testing. The FDA approved four companies to market their antibody kits based on the companies' internal quality control data. The FDA had not independently tested the accuracy of their tests. Some tests were less reliable than they were reported to be.

The CDC and other private institutions were cross-checking various antibody kits to determine the sensitivity and specificity of each testing modality. Until they had such validations, it became challenging for the states to rely on those tests.

Since there was no standardization for the results, it was hard to keep track of more than 100 companies marketing the tests. Do they rely on a test that only detects antibodies, or do they demand tests that measure antibody titers?

ID CARDS OF INTERNATIONAL STANDARD OR STAMPS

Public health officials look from the county, state, and national levels when making policies that affect the population. While over 50% of infected people had recovered from COVID-19 worldwide, it was not clear how long the immunity lasted. From a public health point of view, policymakers cannot check every individual to determine who had and had not recovered, so health policies had to be applied universally.

In fact, some people who recovered from COVID-19 tested positive for a second time in South Korea and China. Hence, there was no full understanding of the extent of immunity.

Some countries considered introducing Immunity Passports like the immunization records presented in schools and workplaces. I used to get tested for TB at work every year, even though I have had a BCG vaccination. Many workplaces strongly encourage their healthcare workers to take 'flu shots, especially those who are dealing with patients directly.

A national ID card to show who had recovered from COVID-19 was a possibility; however, was it the right time to start a movement to push politicians to introduce this policy? Many questions remain unanswered. It is not known if patients who had recovered from COVID-19 and then retested positive for the new coronavirus were infectious. Were these people shedding live viruses, or were COVID-19 remnants such as mRNA chains detected by the tests?

There was growing evidence that some people test positive for weeks after an active infection. There were concerns that tests were being misinterpreted, and people were deemed infectious when they were not. These results kept people from returning to work or reuniting with their loved ones. The term rt-PCR stands for a reverse transcription-polymerase chain reaction. It detects the viral footprints like their RNA. Detecting RNA does not mean a virus is alive or contagious.

South Korea studied 285 COVID-19 patients who had recovered and later tested positive for the virus again. They were not able to replicate those viruses in culture media. That raised the question of whether the patients were shedding infectious viral particles, adding more confusion to the equation.

Based on these observations, Korean officials no longer require recovered patients to test negative for the virus before leaving isolation or returning to public spaces.

Both the WHO and the CDC considered people recovered and non-infectious ten days after their symptoms began, if they had been symptom-free for three days.

CDC Symptom-Based Strategy: If the patient had no symptoms by day 7, then by day 10, they considered them cured. If they became symptom-free on day 12, they considered them cured at day 15.

CDC Test-Based Strategy: They required asymptomatic persons to quarantine themselves for 7 to 10 days after they tested positive for COVID-19. If in the interim, the patients developed symptoms, they followed the symptom-based protocol.

Some people argued that the CDC recommendation of a 14-day quarantine for someone exposed to a person with an active infection might not be in line with these recommendations for those who already tested positive with no symptoms. Some patients could shed viruses for 3-6 weeks after acquiring COVID-19. Hence, the standard guidelines might not prevent all incidences of viral spread. The hope was that people who had recovered might not shed a large viral load.

Again, the current tests that detected RNA remnants related to the coronavirus would not answer the fundamental question regarding whether patients remaining positive were still spreading the live virus.

At least 14 sailors aboard the USS Theodore Roosevelt tested positive for a second time for COVID-19. They isolated the sailors for at least 14 days.

COVID-19 IMMUNITY PASSPORTS?

Immunity Passports are based on the concept that people who had the COVID-19 infection and recovered would have developed antibodies to the virus and, thus, were less likely to carry spread it to others or get a second bout of infection. Having immunity was thought to help people to return to work or travel during an all-out second or third lockdown. These were thoughtful speculations. However, there was no scientific evidence on how long immunity lasted, or how well it protected the person against future COVID-19 infection. So, was the immunity passport concept based on sound reality or proven scientific data?

If vaccination records were acceptable, why not an Immune Passport from a reputable state or federal agency? When a school gets a vaccination record, do they question the name of the doctor who signed the document?

Interestingly, dating services also were interested in the concept, so they could say their member was COVID-19 proof! Many other places like the gym, spa, and beauty salons were also interested in adopting the Immunity Passport approach.

Would Immunity Passports help travelers bypass the compulsory 14-day quarantine when they enter another country?

There were phone apps in the works to document COVID-19 immunity. These apps were mushrooming as fast as the COVID-19 was spreading around the world.

Such apps might be of interest to the hospitality and service industries to make sure employees were not at risk of passing on the infection to clients, and some hotel chains accepted Immunity Passports via apps such as the Onfido app.

But Immunity Passports can create a divisive society. Widespread use might lead to a multi-tier society and discrimination. Some fear people could buy these certificates and that they could create an underground business, fooling not only the people but also the public at large into a false sense of security.

There is also a risk that Immunity Passports might become an additional qualification when choosing new employees or candidates: "She was fine! She was well educated, experienced, and she was COVID-19-proof."

Does the whole concept of Immunity Passports violate an individual's privacy?

Could waitstaff wear a button that read, "I have COVID-19 antibodies!"

In China, people use color-coded QR codes on their smartphones carrying their health information. Shoppers scanned the QR codes at the mall entrances and public places. If they had an active infection, it would prevent them from entering any such public places.

The Robert Koch Institute, the German disease control and prevention agency, was one of a few in the world conducting large-scale random antibody testing. Those results might reveal how rampant the COVID-19 infection was in their country, what percentage of people have antibodies, how much protection those antibodies provide, and for how long the protection lasts. Germany might well be writing the antibody chapter on COVID-19 before the COVID-19 pandemic is over.

A more fundamental question is: how reliable are these antibody tests? Which one should you use? Do you look at tests that simply show the presence of antibodies, or do you want to test for antibody levels to make sure they meet the criteria for protection?

OPPOSITION POINTS OF VIEW

A false-positive test might give false hope. It increases infection risk if people do not take the usual precautions.

Some authorities felt that Immunity Passports were a terrible idea. In the 19th century, one group of people discriminated against others in New Orleans based on whether they were acclimated (overcome by yellow fever) or unacclimated (who did not have the disease). It created a class war where immunity concentrated power in elite hands.

Even at present, you cannot travel to many African countries without a verified yellow fever vaccination.

The WHO has raised concerns that antibodies do not equate to complete immunity. In April 2020, the WHO cautioned against issuing Immunity

Passports, stating that there was no evidence that people who had recovered from COVID-19 had antibodies and were protected from a second COVID-19 infection.

Could Immunity Passports promote unwanted COVID-19 parties where people might deliberately get infected to develop antibodies so they could get a certificate and do whatever they feel like?

Ref:

https://www.cdc.gov/coronavirus/2019-ncov/hcp/disposition-in-home-patients.html

https://www.beckershospitalreview.com/infection-control/recovered-covid-19-patients-who-retest-positive-aren-t-infectious-study-finds.html

https://www.statnews.com/2020/06/08/viral-shedding-covid19-pcr-montreal-baby/

https://onlinelibrary.wiley.com/doi/full/10.1002/jmv.26114

https://www.bbc.com/news/business-53082917

go.nature.com/3cutjqz/

https://www.qr-code-generator.com/blog/qr-codes-coronavirus-china/

Nik Nikam
June 25 at 9:17 PM · 🐾 transmission

FACE MASKS, AND MASK MYTHS UNMASKED!
NIK NIKAM, MD, MHA. HOUSTON, TX
COVID-19 UPDATE

There has been plenty of debate about wearing masks to prevent the spread of coronavirus. At the start of this pandemic, we didn't really have much scientific evidence. And what little evidence we had involved other diseases like the flu, SARS, MERS, which can't be applied to the COVID-19 virus.

The first purpose is to prevent a person with coronavirus infection from spreading the virus to other ...
See More

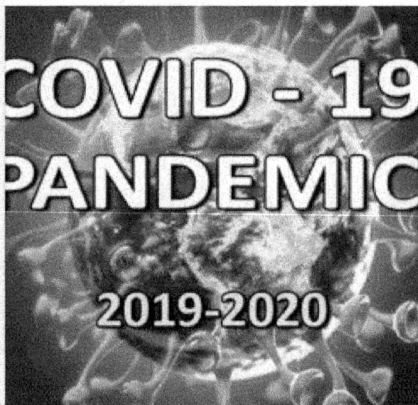

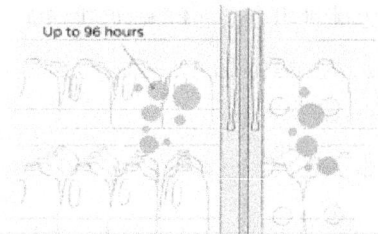

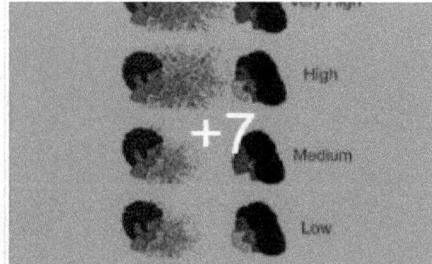

👍❤ Priya Saini Verma and 248 others   67 Comments

17 FLATTENING A CURVE? AT WHAT COST?

The primary goal of the lockdown was to flatten the curve so the states could handle the onslaught of new cases flooding the hospitals and depleting the limited healthcare resources, which could have increased the mortality due to lack of adequate medical care.

People raised many questions about the rush to flatten the curve.

Did the flattening of the curve reduce the overall mortality related to the pandemic?

Did that mean you were stretching the rubber band, but the real end-result was the same?

You might reduce a few cases of death from the pandemic, but what about the social, financial, and economic toll it would have on society?

Look at the figure above. The black line at the bottom shows the health care resources in terms of beds available in New York at present and the bed capacity needed during the peak season. In March, New York had had a bed capacity of 21,200. Based on the model, the demand would surge to 34,500 during the following 4-6 weeks. With the progression of the pandemic, the demand for inpatient beds would increase to 43,600. (Figure 17-1)

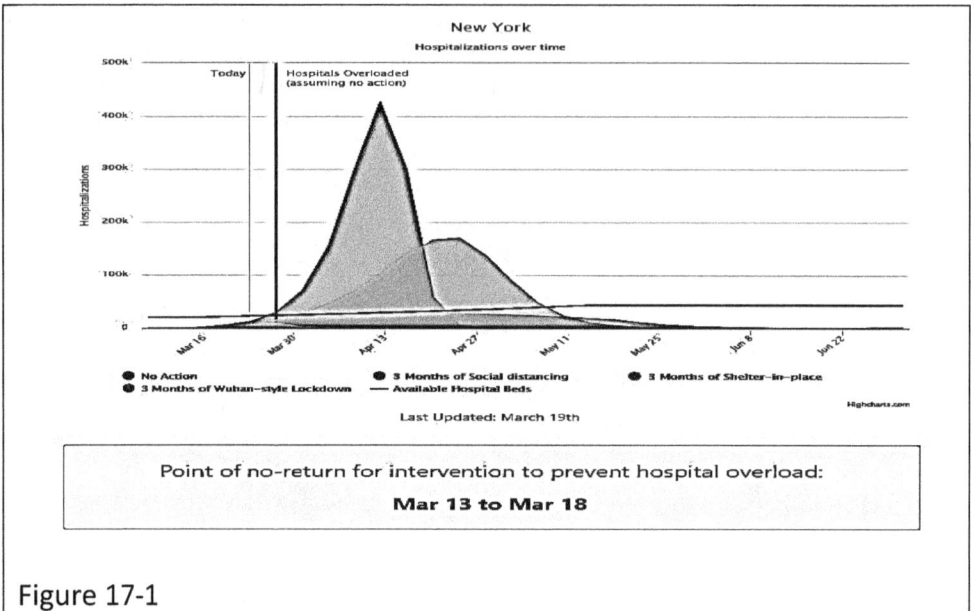

Figure 17-1

When medical resources fall short of demands, it leads to confusion, panic, and uncertainty.

Given New York's medical resources, the only way it would have been able to handle the crisis was for them to follow three months of Wuhan-style lockdown. The tiny cyan color curve represented that at the beginning of the graph on the left side. But they did not have three months' notice, and it would have been impossible to establish three months of Wuhan-style lockdown.

What about the psychological and long-term effects of isolation? Extended lockdowns could lead to depression, stress, anxiety, and more suicides.

What about the fate of people on hourly wages who could not work for weeks or months?

What about the impact of the excess demand placed on the healthcare resources that could not meet the demands that could lead to a collapse of the Healthcare system? It might lead to catastrophic deaths of not only the people who were infected with COVID-19 but others in need of medical care, along with putting a substantial number of healthcare workers at grave risk.

Facing a rapidly raging viral spread creating death, devastation, and desperation along its path, how could governments sit and do nothing?

Could they have said the ultimate results would not be not much different, so go on with your work?

What would you do if you were a leader of the task force in charge of a pandemic?

This was not the first pandemic the world had faced! What lessons did you learn from the past pandemics that you could have applied here?

LOCKDOWN SIDE EFFECTS

Did the lockdown cut the spread of infections to other people? The media and the public questioned whether the lockdown reduced the number of new cases or mortality.

Following the lockdown, New York City reported that over 60% of its new admissions to the hospital came from people in a lockdown situation.

That raised the question of how well the lockdown prevented COVID-19 spread. Living in a high-rise apartment complex was very much like living in a closed environment, such as a nursing home. It exposed people to visitors, common hallways, central air conditioning, elevators, etc. What was the alternative?

We saw people opposing the lockdown and taking to the streets in mass protests. Well, they did that at their own risk. We knew what happened in Sweden, where they had extremely limited mitigation. Sweden had a remarkably high death rate per million compared with most other European countries.

The policies and regulations in different states became a reflection of different ideologies based on political alliances. That created friction between the administration and the people of their states.

More than 40% of the deaths in New York came from nursing homes, where over 50% of residents tested positive for the coronavirus.

Later, we learned that coexisting conditions such as age >70 years, heart disease, hypertension, lung problems, diabetes, and obesity all played a significant role in the increased number of cases and deaths in the US.

IS A SUSTAINED QUARANTINE THE RIGHT APPROACH?

Just because quarantine was lifted, it did not mean the virus had left town.

More people were expected to get sick, with 10-20% likely to need emergency department care.

There were two ways for people to get immunity to the coronavirus. One, catch the virus, hope you do not get too sick, and get over the virus with mild symptoms. The antibodies following the infection would provide some protection. The second option was a vaccine that does the same thing by fooling the body to think it has the virus, so the body produces antibodies to fight the real virus when exposed.

Based on these facts, they were expecting a bumpy ride for the next several months. At the time of this writing, they were still in search of a safe way to gain herd immunity.

People expected the summer months to slow COVID-19 spread and hoped the virus might very well disappear in a few months.

They also were expecting a resurgence of COVID-19 during the colder winter months along with the seasonal flu. That could overburden the healthcare system again

It was hard to speculate what was right. People were told to stay home. They stayed home, felt safe, and protected. However, the prolonged lockdown took a toll on people's minds, and some ignored all self-mitigation rules.

When New York reported that 60% of fresh cases came from people at home during the lockdown, people questioned where the infection came from. Was the virus circulating in the air? Did they catch the viruses from carts in the grocery store? Did the food delivery bags carry the virus? Did they get infected when they went for a walk in the park?

If sixty percent of the cases in New York came from people sitting at home, why did they need a lockdown? They could have gone to work and applied the same self-mitigation rules at work.

The best infection indicator numbers to follow are:

- A daily number of new admissions to the hospital, region by region
- Number of total COVID-19 patients in the hospitals, region by region, daily

Death was not as reliable a number as the number of new admissions for the allocation of healthcare resources. The death numbers could be under-or over-reported or delayed, etc.

LOCKDOWN AND SELF-QUARANTINE PROS AND CONS

Did lockdown and quarantine help to cut the spread of infection to other people? New York City reported that more than 60% of hospital admissions were from people in a lockdown situation.

This raised the question of how well lockdowns prevent the spread of COVID-19 infection. More than 45 countries issued lockdowns in March and April except for Sweden. Sweden encouraged people to maintain social distance, but their schools, restaurants, and businesses stayed open. Even though Sweden was not at the top of the list for total deaths per million, it was close to the top on March 16.

It was difficult to convince people it made sense to close New York City for a handful of deaths from coronavirus. Gov. Cuomo had said he would not approve New York City mayor's lockdown order for the City.

It was during this time the virus was rapidly spreading from infected people to thousands of people in densely populated buildings and workplaces.

Gov. Cuomo ordered the lockdown of New York State on March 22, 2020. That was 21 days after the first case report. Lockdown works best when instituted early in the epidemic phase and is best when just a handful of cases have been reported. That is what the leaders in Taiwan and South Korea did.

They blocked all incoming flights to prevent infected people from carrying the disease into the country.

Self-mitigation should start from day one at ground zero. The best way to beat any viral epidemic or slow viral spread is early self-mitigation. But for COVID-19, there were no guidelines from any major source on what steps people needed to take, like the ones you know now, such as social distancing, washing hands, wearing masks, and avoiding meetings and crowded places.

Nik Nikam
20 hrs

HOUSTON, TEXAS, RACES TO HOTSPOT-ONE FOR COVID-19 INFECTIONS POST LOCKDOWN! – AN EYE OPENER FOR THE REST OF THE COUNTRY

Nik Nikam, MD, MHA. Houston, TEXAS

AN UPTICK IN CORONAVIRUS CASES POST-LOCKDOWN
... See More

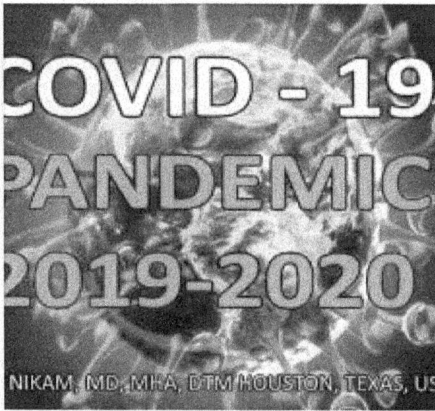

Houston region sees rise in coronavirus cases

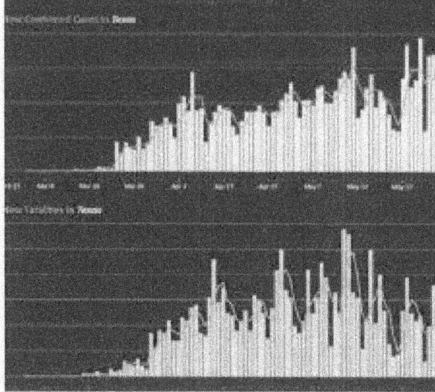

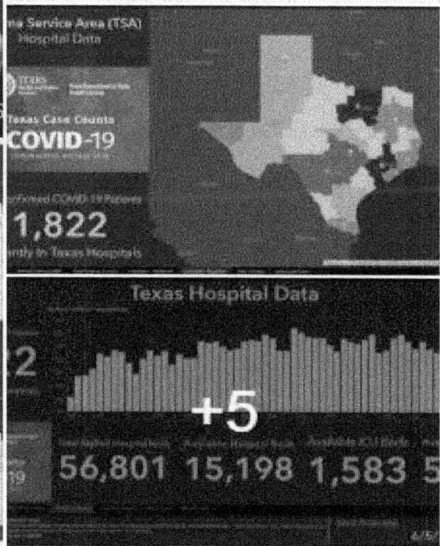

😮😢 57   106 Comments

👍 Like   💬 Comment

View 36 more comments

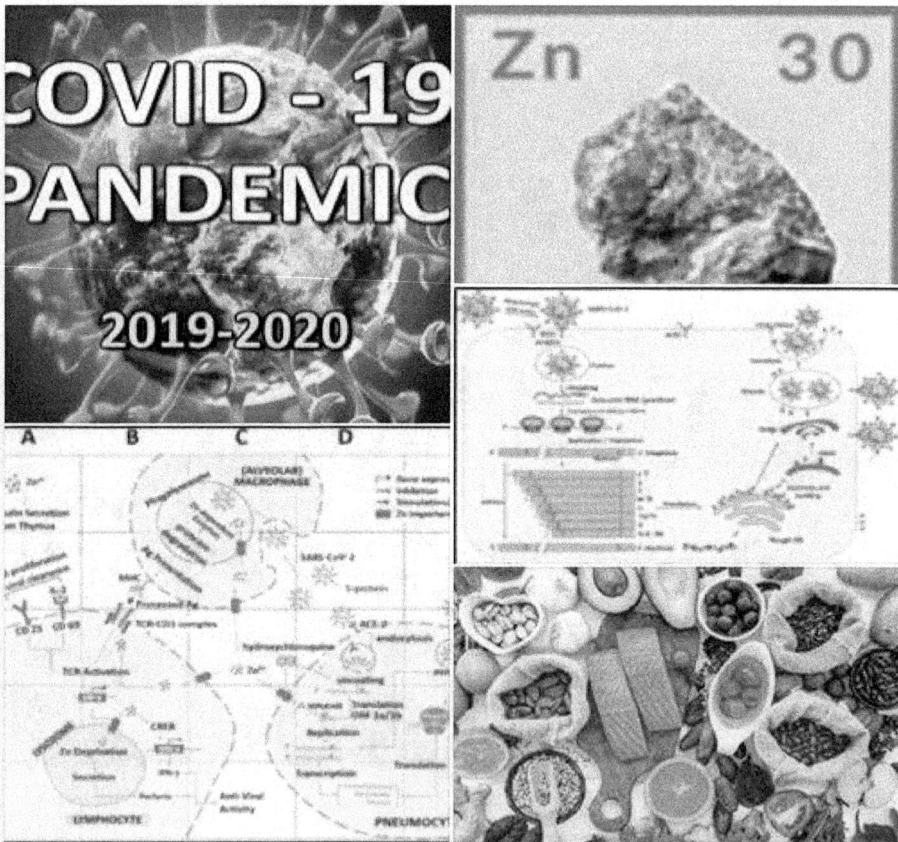

Nik Nikam
July 22 · 🏷 treatment recs

ROLE OF ZINC IN COVID-19 PATIENTS?

Nik Nikam, **MD, MHA, HOUSTON, TX**

Coronavirus infection has no know treatment. This has attracted a lot of interest in exploring various antiviral agents and immunomodulators to prevent the coronavirus replication and minimize the tissue damage. Zinc is one such element that has antiviral properties. They have tried Zinc (Zinc) in many viral infections such as hepatitis C, HIV, and others. Here, we study its role in the management of Covid-19 pa...
See More

00 78   174 Comments

18 EMOTIONAL & PSYCHOLOGICAL CHALLENGES

The COVID-19 pandemic, which started in December 2019, had affected more than 200 countries, raising concerns of widespread panic, and increasing anxiety in individuals subjected to the real or perceived threat of the virus. The pandemic was more than a mere medical phenomenon. It affected individuals and societies at many levels. Stigma and xenophobia were two aspects of the societal impact of the pandemic outbreak (WHO).

Medical mismanagement, time delays, misinformation, and political finger-pointing all added to the complexity of emotional problems and mistrust. These medical mistrusts can cause some people to defy mitigation guidelines, accounting for a resurgence in the number of new cases, thus, becoming a vicious cycle that threatens core belief in the medical system, leading to more stress, anxiety, and exhaustion.

The protracted COVID-19 ordeal, with no end in sight and an ever-gloomier outlook daily, had the potential to become a perfect storm of social and psychological illnesses that could stress an already-fragile mind and lead to mental breakdowns.

It was the uncertainty, the isolation, the restrictions, the constant bombardment of the "Doomsday News" by the media every hour, on the hour that polluted our minds and hampered our ability to cope with the reality and the limitations forced upon us.

School closings affected over 30 million students, turning their parents into full-time caregivers and homeschool teachers. Now, you can appreciate why parents were becoming increasingly irritable.

The added stress of total isolation, physical distancing, and routine use of masks could be demoralizing and heighten responses to a stressful situation. In a poll conducted by the Kaiser Family Foundation in April, over 50% of adults reported worry or stress due to the pandemic affecting their mental health and wellbeing.

How do you live through these pandemic cycles and still manage your mental, physical, and emotional lives to live another day to say, "I did it"?

How you respond to the COVID-19 pandemic depends on your background, support from family and friends, financial situation, health and emotional stability, and community involvement, among many other factors.

None of us were mental health specialists, yet many of us faced these challenges. It was a good time to recognize many common psychological and social challenges we faced and learn how to deal with them.

I am the first to admit that I am not a mental health specialist. I am a lifelong student of art and medical science. My day job is cardiology, and my passion is medical journalism and social media communication. A few years ago, I authored a book on the stress-less lifestyle.

COMMON EMOTIONAL AND PSYCHOSOCIAL PROBLEMS

- ➢ Sleep disturbances
- ➢ Eating disorder, mostly overindulgence
- ➢ Difficulty in concentrating
- ➢ Worsening of chronic medical conditions
- ➢ Exacerbation of mental illnesses: anxiety, depression, etc.
- ➢ New habits like smoking, drugs, alcohol
- ➢ Emotional outbursts of anger, frustration, restlessness

> ➢ Domestic violence and suicides
> ➢ Financial instability

Ralph Waldo Emerson said that what lies in front of us pales in significance compared with what lies within us. I would like to add to that what lies in front of us (COVID-19 challenges) pales in significance compared with what lies ahead of us (our overall future). This entire chapter focuses on "what lies within YOU" and how you can empower yourself.

It is not what is happening to you, but how you respond to what is happening to you that determines your success and outcomes. Two individuals facing similar situations might see things differently and respond according to different emotions and outcomes.

You need to focus on solutions more than the problems that have created that turmoil in your life. When you focus on the solutions, your mind is engaged constructively and assertively, so you can control your emotions, reactions, and outcomes.

HIGH-RISK GROUPS

- Predisposition to a higher risk for severe illness from COVID-19, including people with underlying health conditions
- Children and teens, people caring for family members or loved ones
- Frontline workers such as health care providers and first responders, retail clerks, and essential workers who work in the food industry
- People who have existing mental health conditions, and those who use substances or have a substance use disorder
- People who have lost their jobs had their work hours reduced or had other major changes to their employment

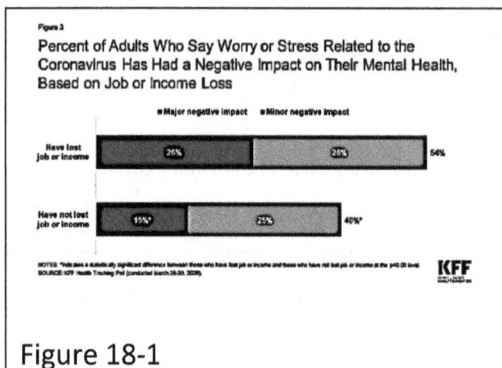

Figure 18-1

- People who have disabilities or developmental delays, those socially isolated from others, people who live alone, and people in rural or frontier areas were more vulnerable. (Figure 18-1)
- People in some racial and ethnic minority groups who do not have access to information in their primary language.
- People experiencing homelessness.
- People who lived in a congregation (group) setting, who needed to care for their community and themselves. (Figure 18-2)

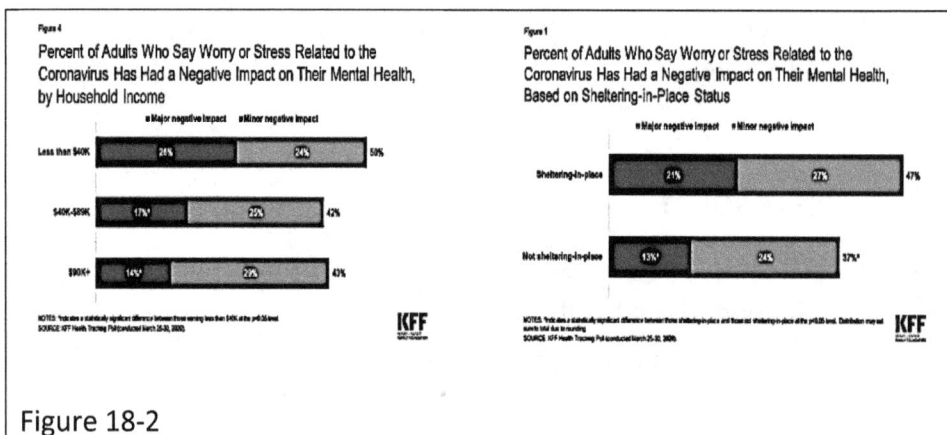

Figure 18-2

LESSONS FROM PRISONERS OF WAR

A prisoner of war confined to a dark cell, goes through an initial shock and disbelief. Then anger and frustration follow. Next, the reconciliation and acceptance of reality. He tries to find a routine and peace despite spending years and sometimes decades confined. He tries to look for the inner strengths within him and build on what he had learned as a child, be it rhymes, poetry, or psalms. He might even try to find peace with nature and try to live a meaningful life. We can take those lessons and make our lives more meaningful to us and others in a time of crisis such as the COVID-19 pandemic!

ROAD TO RECOVERY

When stress overwhelms you to the maximum, it is hard to take care of your family and friends. But getting engaged in helping others could take your mind off the boredom and be subliminally therapeutic. Feeling good about yourself inspires you to help others, even with physical distancing.

FIRST THINGS FIRST

You need to define what is eroding your emotions and draining your life day by day. Write three things that have transformed your life into a living hell. If you do not know what your problems are, you cannot focus on your solutions.

First, you need to identify what you can do about it. You cannot fix the weather, undo an accident, or bring back lost loved ones. But you can address the aftereffects that are taking a toll on your emotions, nerves, and livelihood.

Think of your brain as a box that can only accommodate ten emotions. If negative, desperate, self-destructive thoughts take all the ten spots, there will be no place for good, productive thoughts in your mind to start a journey toward a healing process.

You need to remove thoughts that are not under your control and replace them with new constructive thoughts to focus like a laser beam instead. Invest all your energy, time, and effort on those two new thoughts, so you no longer are feeding the negative thoughts. When you stop feeding negative thoughts, they lose their potency like COVID-19 would if it did not have a steady supply of new hosts to replicate in.

BE SELFISH

I once heard a psychologist present on how vital it was for us to be selfish. I was more curious to know what she was smoking! It turned out the title had a second part, which read "IN ORDER TO HELP OTHERS." For the next ten minutes, I was arguing in my mind this was false advertising. The more I resisted, the more she was dragging my attention into her concept. She said if you are not physically, mentally, and emotionally sound, you cannot help others. So, human services begin with you. When you get the person within you right, the entire world looks wonderful and is a much better place for the people around you!

The process of dealing with COVID-19 challenges begins with you. I will share some tips I have learned along the way and that I have used them in my daily life. Hopefully, these will help you feel good about yourself and inspire others to help more people.

SOCIAL MEDIA QUARANTINE

Social media is one of the greatest sources of misinformation, exacerbation, fake news, and hatred, and it can be downright depressing. It can be dehumanizing. Find a private group that focuses on uplifting your spirits and engages you in an optimistic and constructive manner to replace those negative thoughts in your mind. These are available on many social media forums.

What? You received no 'likes' on your post, and someone else got 937 'likes'? Is that supposed to inspire you and make you feel good about yourself? Do you care about the person who has not liked or commented on a single post of yours? Simply mute their posts for 30 to 60 days (or 30 to 60 years…) or delete them from your friend list as they have not extended any friendship to you. Protect your emotional investment and save it for people who interact with and encourage you, instead of letting people who use social media to showcase their 'perfect' life make you feel bad about yours!

Avoid the pros and cons of any topic on social media or in special groups. Social media is making billions of dollars on your time. You are an unpaid employee of social media, which is exploiting your intelligence, time. Instead, focus on small special-interest groups that provide self-help tips and inspire you to learn more coping skills.

Whatever you do, it must take you from point A to point B, even if your goal is rest and relaxation.

DEVELOP A NORMAL NEW ROUTINE

Create a new timetable in quarantine for filling your time when stuck in the house for 24 hours, seven days a week. Writing is a valuable use of your time. Start with writing one thing on a piece of paper and keep adding something to it each day. If you get to page 70, create a formatted book in MS Word, upload it to Amazon Kindle Books, and start making some nice, soft, green money! Who knows—you could become a bestseller on "How to Live in Solitary Confinement and Still Write a Book!" If you are reading this book, that's how I made it through more than six months of being a prisoner of time!" Writing served as my mental break from my eight hours a day of telemedicine as a cardiologist in solitary confinement.

Enjoy an extra two hours of sleep. Sleep is good for the body and mind. Spend an hour watching videos on how to cook new dishes, learn to play a musical instrument, a new style of art, or anything that fascinates your imagination. If you think you are stressed to the maximum, wait till you watch a cockpit view on how to fly a Boeing 747 jet on YouTube. That should calm your nerves!

Make sure you dress-up and make-up, even to your home office. Make it a habit to stick to your work desk at least a few hours at a stretch. That provides a sense of accomplishment, which propels the happy hormones, namely, endorphins.

Allocate 30-60 minutes for exercise, meditation, prayer, music, or concentration. Turn on YouTube, pick a series, and follow them. You could even produce your own YouTube series on how to cope with quarantine.

Cooking is a great hobby to replace your stress. There are millions of videos on how to prepare your favorite gourmet dinner from start to finish. Yes, you might be the CEO of a billion-dollar company. But if you are stuck alone in the house, a pandemic could not care less about your title. You might learn a fancy recipe or two, which might come in handy to surprise your friends and family once quarantine ends.

Be vigilant when you are walking in a crowded park, especially in areas where the percentage of people testing positive for virus is high, such as in San Antonio in July, where infection rates reached 20%.

Avoid relying on alcohol. Getting hooked on a substance is like overbooking your brain box with increasingly self-destructive negative habits. Try to get rid of your negative thoughts and habits that are eating your brain, robbing your peace and tranquility.

Social distancing does not mean social disconnection. It is time to increase your social connection with positive and uplifting groups of people even if you cannot connect in person. I saw people spending long hours in WhatsApp groups debating political issues and current events, dehumanizing, and tearing each other apart. Engaging in such conversation can be highly corrosive to your mind and psyche. There are better ways to connect.

VIRTUAL GROUPS

Create a small professional, family, or friend groups that can meet on MS Team, WebEx, or Zoom. Talk about things you did that had a positive impact on your life and uplift others. That is using your time constructively in a way that could help other people.

Keep track of one reliable source for information and ignore the other 30 channels with their political biases so cleverly integrated. Look at the raw source data instead. Consider your role in preventing virus spread and how you can accomplish that. The rest has no impact on your life unless you put yourself in the middle of a crowd with no protective gear. Avoiding negative media sources is one way to avoid spending hours unproductively arguing on social media.

AVOIDING SLEEPLESS NIGHTS

Many people experience sleepless nights. Try combat that by keeping yourself busy with positive tasks all day long. When you are tired, you are likely to fall asleep. However, if you have a mission-critical job and you cannot sleep at night because of nightmares, fears, or anxiety, seek medical help for a short-term solution as needed. You should do that earlier rather than later. You could try over-the-counter remedies after consulting with your physician. Sleeplessness was a common challenge for doctors who worked 12-hour days with COVID-19-positive patients. Avoid caffeinated drinks late in the evening or at night as they will prevent you from sleeping.

YOUR HEALTH COMES FIRST

When things are falling apart in all directions, it might tempt you to let the guard down and ignore your chronic medical problems. Remember what I said—you must be selfish to help others! This means taking care of yourself to the best of your ability and rewarding yourself from time to time for virtuous deeds so that your mental health is sufficiently sustained to better assist others.

Engage in a routine exercise program within your confines. When you exercise for 20 minutes, you feel good for the rest of the day. Eat healthy foods. If you have too much free time, take the time to cook a special dinner and get your family members involved in the preparations.

SUICIDE PREVENTION

Suicide risk is higher among people who have experienced violence, including child abuse, bullying, or sexual violence. A sense of isolation, depression, anxiety, and other emotional or financial stresses might raise suicide risk. Do not let a pandemic become the final trigger.

Many times, after a suicide, even the person's best friends comment they did not know that their friend was planning such a thing. But, when you look back in time, there are often clues on a person's social media, their timing, their actions, their absence from activities and family events, etc. Some might already be on treatment for mental illnesses like panic attacks, depression, etc.

Unusual activities or no activities in an individual should be a clue, and intervening can save lives.

There are ways to protect against suicidal thoughts and behaviors—for example, support from family and community to help the person feel connected. Additionally, in-person or virtual counseling or therapy could help with suicidal thoughts and behavior, particularly during a crisis like the COVID-19 pandemic.

The mental health system in the US is very fragmented and difficult to access, with state license restrictions and prohibitive costs. It is grossly under-funded to deal with millions of people with a variety of mental problems. A federal emergency hotline for people in emotional distress registered a more than 1,000 percent increase in April compared with the same time last year. Over 50% of people reported that the COVID-19 pandemic had influenced their emotions. Similarly, various other mental health hotlines have had a substantial increase in the number of calls during the past three months. Sadly, the federal government was not even talking about it.

The suicides of two New York healthcare workers underscored the reality of the emotional and psychological stress arising from dealing with deathly sick COVID-19 patients for whom they could offer extraordinarily little treatment. Lorna Breen, a top New York emergency room doctor with no prior history of mental illness, struggled with the increasing emotional burden of the COVID-19 pandemic before she committed suicide. Similarly, a Bronx emergency medical technician took his life.

Based on data collected after natural disasters, terrorist attacks, and economic downturns, researchers have shown an increase in suicides, overdose deaths, and substance use disorders in unprecedented stressful situations.

LACK OF ADEQUATE MENTAL HEALTH PROFESSIONALS

Because of a lack of adequate funding for the mental health issues related to the COVID-19 pandemic and lockdowns, many existing therapists and psychologists had to shut their businesses. Additionally, they often could not reach people remotely due to state license restrictions, rules, and regulations. That was reflected in workers' productivity, patient health, and the economy.

As the State and Federal governments focused on reducing the number of patients in the hospital with COVID-19 infections, they also needed to allocate funds for delivering mental health services online. Mental health symptoms might trail by weeks or months after the beginning of a pandemic. Increasing unemployment numbers show a proportional increase in suicide rates. Meadows Mental Health Policy Institute, a Texas nonprofit, created models that suggested if the unemployment raised to 20 percent, like the levels recorded during the 1930s Great Depression, suicides could increase by 18,000 and overdose deaths by more than 22,000.

A study of 1,257 doctors and nurses in China during its country's COVID-19 peak found that 50 percent reported depression, 45 percent anxiety, and 34 percent insomnia.

In early April, mental health organizations estimated they needed $38.5 billion to save treatment providers and centers and another $10 billion more to respond to the COVID-19 pandemic. They received less than 1% of the requested amount—$425 million in emergency funding and $375 million to state and local organizations.

Health care providers were particularly vulnerable to emotional distress given their exposure risk to the virus, concern about infecting and caring for their family members, shortages of personal protective equipment (PPE), long working hours, and emotional entanglement in the ethically contested resource-allocation decisions.

State medical associations should provide education and training on psychosocial issues related to a crisis to healthcare leaders, first responders, and healthcare professionals. They could accomplish that through state medical associations and mental health departments as part of their continuing education series.

Healthcare systems need to address the stress on individual providers and general operations by monitoring responses and performances, altering assignments and schedules, modifying expectations, and offering much needed psychosocial support.

IMPORTANT CONTACT NUMBERS

Some important number you can keep handy and pass on to your family and friends:

- Call 911
- Distress Help: 1-800-985-5990
- National Suicide Prevention: 1-800-273-8255
- National Domestic Violence Hotline: 1-800-799-7233
- National Child Abuse Hotline: 1-800-422-4453
- National Sexual Assault Hotline: 1-800-656-4673
- The Eldercare Locator: 1-800-677-1116
- Veteran's Crisis Line: 1-800-273-8255

Find a health care provider or treatment for substance use disorder and mental health:

SAMHSA's National Helpline: 1-800-662-4357

Ref:

who.int

cdc.gov

https://www.washingtonpost.com/health/2020/05/04/mental-health-coronavirus/

https://www.nejm.org/doi/full/10.1056/NEJMp2008017

https://www.kff.org/coronavirus-covid-19/issue-brief/the-implications-of-covid-19-for-mental-health-and-substance-use/

https://www.kff.org/coronavirus-covid-19/issue-brief/the-implications-of-covid-19-for-mental-health-and-substance-use/view/footnotes/

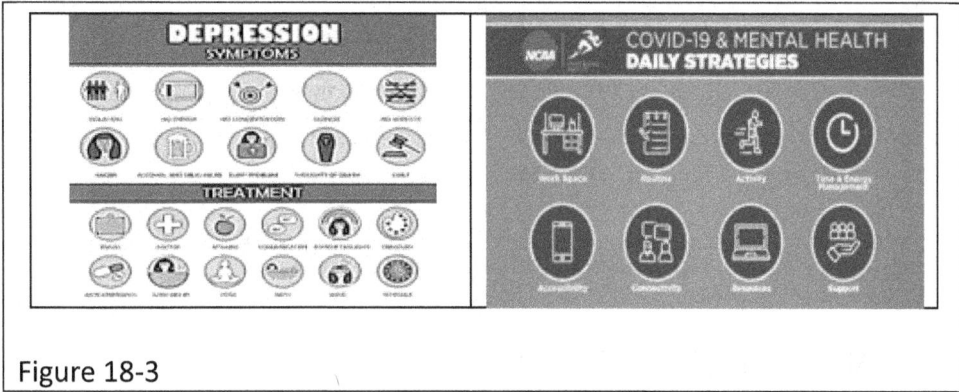

Figure 18-3

Nik Nikam
Yesterday at 4:11 PM

LIFE TRANSFORMATION IN 60 DAYS FROM CORONAVIRUS

Nik Nikam, MD, MHA. HOUSTON, TEXAS

Who would have ever imagined in February 2020, that America will come to a standstill, as the coronavirus pandemic swept the world like a Tsunami?

The first cases were reported in New York city in the first week of March 2020.

As of May 15, 2020, there are more than 1.5 million cases of active coronavirus infections in the United States, accounting for more than 85,000 deaths and still there is no ...
See More

United States
Coronavirus Cases:
1,485,912
Deaths:
88,550

Texas
Coronavirus Cases:
46,787
Deaths:
1,308

Coronavirus Cases:
4,660,658
Deaths:
309,710
Recovered:
1,777,303

+14

Nandana Kansra and 142 others   87 Comments

Like   Comment

19 TRAVEL CHALLENGES–COVID-19 PANDEMIC

IS TRAVEL WORTH IT?

You thought it was time to shake that cabin fever, get fresh air, move, and meet some old friends? Think again, twice, three times! You do not know what surprise is waiting for you on the road to unfounded happiness. If you were unhappy at your own home, wait till you get on the road. The road conditions might be fine, but the weather could play havoc on you. It might be too hot, or it might rain on your special parade.

Kids are yelling in the backseat because they want to use the restroom. You cannot stop anywhere to freshen up. Anywhere you stop, you could be exposed to other people. You might have to eat food from places you would not normally eat. You will not know if the people who handled your food are infected. The restrooms might be full of coronavirus.

With all those worries, with the kids yelling and screaming in the back seat, your dozy eyes half shut, the car tires swing from one edge of the road to the other. Suddenly, your spouse's scream saves your life from an oncoming car. You wonder whether you would have been better off on your lazy-boy chair with a cup of tea and your iPad. How can anyone compare a long-awaited vacation to a lazy-boy? I bet I could! But my vote doesn't count!

As you enter the city, it might look like a ghost town. Most of your favorite shops are closed. Even the restaurants are closed. The option is takeout.

Finally, you reach your friend's house.

Then, your friend or family member might be hesitant to receive you as a guest. They might fake a smile, but their feelings tell a different story. They may be in no mood to sit and have a wonderful conversation with a carload of uninvited guests on top of their job, cabin fever stress, homeschooling three kids for 8-hours a day, cooking four times a day, cleaning, and dealing with others' temper tantrums!

If you dare to visit any friends or relatives uninvited, the first question they ask will probably be, "when are you leaving?" They have their hands full and did not need guests to add fuel to their family feud. If you do visit them, then you must live by their house guidelines. You should know their routine. This is not a motel.

Could you imagine their kids and your kids running rampant in the streets, and you do not even know if the neighborhood is safe.

If you cough or sneeze due to your allergies or the change in weather, be ready to answer a series of interrogative questions such as, "Why are you coughing? How long have you been coughing? Are you taking any medicines? Did you get tested?"

After your first cough or sneeze, they may try to avoid you like the plague. Not because they want to ignore or isolate you, but they value their lives more than your friendship.

You sit for dinner, but your kids hate their food. Your kids throw a fit because you didn't care to stop and pick up their fast food, so they ruin your evening and dinner!

Finally, you settle down for that intimate one-on-one time with your friend. You had to catch-up on a lot of things like not being able to go to a salon, take care of your nails, etc. The restless kids drive you and your host to total insanity. Finally, you whip up a plan to put them to bed. The kids want to sleep together. Then, you hear the screams from the bedroom: "Mommy, it's hot here. Mommy, it's too cold, it is freezing. I am shivering."

You yell at your kids, the host yells at her kids, to no avail. And the kids join you in sync.

Finally, you cut short your intimate conversation and try to get some relaxation. You realize it is not a hotel room where you can control the temperature but a house with central heat and air conditioning with the host's mood deciding the temperature and humidity.

Early in the morning, you get the car engine warmed-up and ready to head home. You stuff the kids in the back seat with a warm blanket wrapped around them and head home without a word of conversation.

"Do you want a cup of coffee?" You try to interrupt the constant engine whine.

"Did I ask you? I want nothing?"

"If you want to freshen up, let me know? I can stop the car at the next rest area."

"Will you shut up and keep your eyes on the road! Quit staring at me!"

Three hours go by, not a single word, no music, no noise except for the hissing from the overheated engine. And occasional snoring from the kids.

Two hours later, you say, "Kids, we are home!"

"For God's sake, don't yell! Don't you see the kids are asleep?"

"Thank God!" yelled the kids

A few minutes later, "My head is splitting. John, could you make that coffee for me?"

"Yes, Ma'am, right away! Cream and sugar?"

"What did you say?"

"The coffee is almost ready!" John tried to salvage the conversation.

"You know, there is nothing like home, sweet home!"

"Yes, Ma'am! I agree with you 100%. I will promise you I will never leave home again!"

"Enough of that melodrama. Where is that ointment! I have this splitting headache. "

" Me too! "

"From doing what?"

20 DIFFERENT PRESENTATIONS AND RESPONSES

In the next several chapters, I explore how several countries responded to the COVID-19 pandemic, what challenges they faced, how they overcame them, and what their outcomes were. I conclude with points that could form the foundation for dealing with the next pandemic to save lives, contain the virus spread, and preserve the economy.

I cover the following countries, some from the Asian continent, some from the European block, and some from other parts of the world:

- The United Kingdom
- Germany
- Sweden
- The United States
- New Zealand
- Japan
- South Korea
- Singapore
- Taiwan
- China
- India

WORLD·LEADERS·IN·2020¶

DONALD·TRUMP·-·US¤

ANGELA·MERKEL·GER¤

XI·CHINA¤

BORIS·JOHNSON·UK¤

GUISEPPE·ITALY¤

STEFAN·OEVEN·SWEDEN¤

SHINZO·ABE·JAPAN¤

NARENDRA·MODI·INDIA¤

MOON·JOE·IN·S·KOREA¤

TSI·ING·WIN·TAIWAN¤

JACINTA·ARDERN·NEZ·ZEAL¤

ERDOGAN·TURKEY¤

21 NEW YORK: THE UNITED STATES' EPICENTER

At the beginning of March, New York City and the New York state became the epicenter of COVID-19 in the US. Washington state had reported a few cases in the middle of January 2020. New York gripped the nation and the world as the COVID-19 cases shot through the stratosphere like never before. (Figure 21-1)

As of September 11, the New York state had more than 476,000 cases, far more than the number of cases reported in the entirety of China. The total death toll stood at 33,109. That was 9–10 times the number of deaths seen in Wuhan, China.

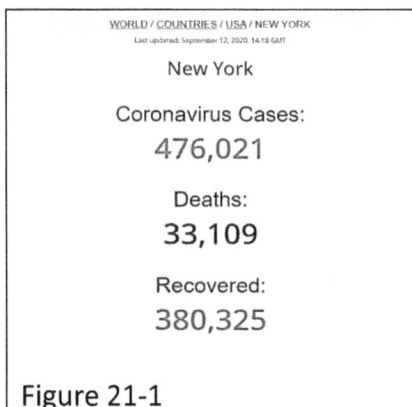

WORLD / COUNTRIES / USA / NEW YORK
Last updated: September 12, 2020, 14:18 GMT

New York

Coronavirus Cases:
476,021

Deaths:
33,109

Recovered:
380,325

Figure 21-1

Since COVID-19 broke out in Wuhan in January, more than two million people had traveled to New York City, the biggest hub for international air travelers. The US government blocked all flights coming from China on February 1. By that time, tens of thousands of people had already traveled back to the US after the Chinese holidays.

The state of New York had a population of 19 million, with a population density of 27,100 people per square mile. New York City (NYC), which was the most densely populated place in the state with people living in small crowded spaces in high-rise complexes, became a perfect medium for the COVID-19 epidemic to spread with a blazing speed like the California forest fires.

NEW YORK COVID-19 PANDEMIC TIMELINE

The state of New York reported its first case on March 12, 2020.

The number of new cases per day at its peak at the end of March topped 10,000. Then, there was a steady decline. By the middle of June, the number of cases per day was in the hundreds.

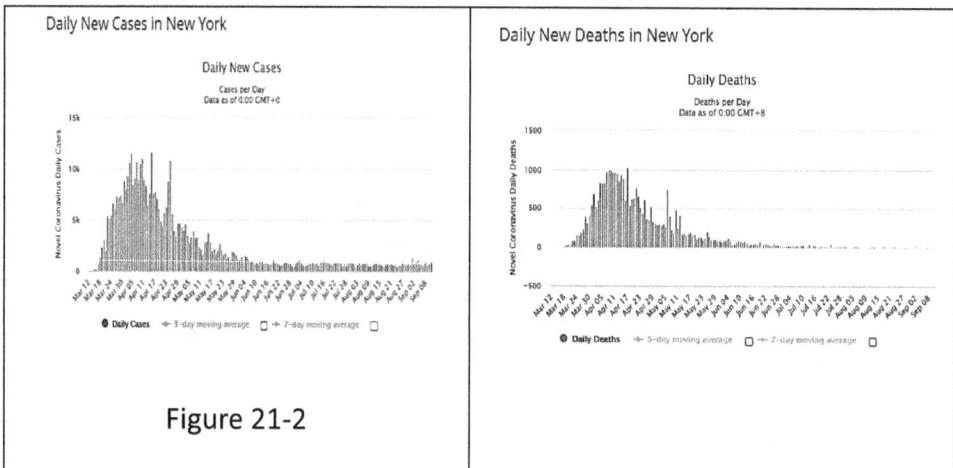

Figure 21-2

NY New Cases 9-11-2020   NY Deaths – 9-11-2020

The death rates peaked at 1000 cases per day in the middle of April, and since then, the numbers have come down and stayed down.

Unlike the southern regions like Florida, Texas, and California, New York did not experience a second spike. They remained in lockdown during July and August, and the actual reasons were unclear as to why New York and New Jersey did not experience a second spike despite the strikes, riots, etc.

New York instituted lockdown on March 22, at which time the state was reporting more than 10,000 new cases per day. The best time to institute lockdown would have been before the epidemic took off. During its peak, New York Gov. Cuomo reported they needed more than 50,000 ventilators and an additional 10,000 beds to deal with the enormous surge of COVID-19 cases.

A CALL FOR MEDICAL PROFESSIONALS: New York Mayor De Blasio put out a call to retired medical professionals and those displaced from their private practice groups to join the city in providing the much-needed medical help. In fewer than 24 hours, more than 1,700 physicians signed-up for this noble medical mission.

A great lesson they learned was to ask for whatever they were short of, be it medical professionals, supplies, or medicines. As most of the private hospitals and outpatient surgery remained centers, they shared their resources with hospitals with immense pride and commitment.

NEW YORK COVID-19 PANDEMIC TIMELINE

MARCH

The first case of COVID-19 in New York was reported on March 1, 2020. On March 3, a lawyer was admitted for pneumonia and later tested positive for COVID-19. On March 4, eleven more people tested positive. Nine of them had had contact with this person. By April 10, New York had surpassed any other country in case numbers.

By March 6, the NY state had not even ordered the required N95 masks and PPE.

On March 7, Gov. Cuomo declared a state of emergency after the state recorded 89 cases. By March 11, they moved most university classes online.

New York recorded the first COVID-19 related death on March 14—an 82-year-old patient with a pre-existing lung condition. By March 24, there were 25,000 cases and 210 deaths. More than 200 NYPD officers and civilian employees had the COVID-19 virus. It also affected 2,774 NYPD employees, representing 7.6% of the workforce.

They were running short of ventilators, and the ICU beds were filling up with COVID-19 patients with respiratory failure, forcing the hospitals to split ventilators between two patients as of March 26. Many patients needed ventilators for 11 to 21 days, as compared with the usual patients who required them for 3-4 days.

Healthcare professionals described the scene at Elmhurst hospital as apocalyptic. On March 25, they had 13 deaths in 24 hours.

New York set up a 68 bed COVID-19 ICU unit in Central Park to accommodate the overflow of sick patients.

Testing was slow, which created a bottleneck. Often, it took more than a week to get the results.

On April 16, Gov. Cuomo ordered all New York residents to wear masks. New York, New Jersey, and Connecticut started a contact tracing program.

The central focus during the early part of March was the lack of N95 masks, PPE, and ventilators. Some of this was due to a lack of preparedness, no reserved stocks for epidemics, supply channel issues, along with delays in procuring the supplies.

By the middle of March, the number of cases in New York surpassed the total number of cases in many small countries. Yet, New York was lagging California in implementing a more drastic measure like the "Shelter-in-Place" order introduced by California on March 19, 2020. New York needed to restrict non-essential movement. Any amount of medical supplies and personnel would not be adequate if they did not cut COVID-19 spread and prevent further escalation of case numbers.

On March 28, the US had surpassed more than 100,000 cases, and New York accounted for over 50% of them. The US epidemic was about 10-12 days behind the Italian epidemic.

The US also was doing more COVID-19 tests than any other country. Partly, that could have accounted for increased numbers. The percentage of positive cases among those tested were also higher compared to those from countries like Germany or France.

Likewise, more than 100 people from the police department had tested positive for the coronavirus

By March 30, New York had 66,497 active cases and 5818 deaths.

At its peak in the middle of April, New York was in a steep upslope of the bell curve, perhaps the steepest compared to that of most other countries. It might mean the spread was very virulent, or that they were testing more people compared to that of other countries, or it may have been a result of the population size and density.

APRIL

On April 6, statewide stay-at-home order and school closures extended to April 29, and later extended to April 29. On April 15, they ordered Face masks/coverings in public places where social distancing is not possible.

During the peak period in April, New York had more than 10,000 cases per day.

New York City Mayor DeBlasio emphasized they would run out their supplies by April, and they wanted to Federal Government to purchase and provide three million N95 masks, 50 million surgical masks, 45 million surgical gown, and 15,000 ventilators.

In April, President Trump invited major company executives to take the lead role in manufacturing the ventilators that were so desperately needed to serve an ever-increasing number of cases of acute respiratory failure from the COVID-19 infections. By the end of April, the US had manufactured 50,000 extra ventilators.

They converted the Javid Center into a makeshift hospital with more than 1000 beds to accommodate fresh cases of COVID-19 cases. But they never used it during the COVID-19 epidemic.

President Trump ordered a Naval Military hospital ship with additional 1500 beds fitted to handle the COVID-19 patients. It, too, remained largely unused, at least to the extent they were intended to be used.

MAY

On May 1, schools were ordered to remain closed for the rest of the semester.

On May 15, they opened four counties with the least risk of viral spread. May 23, they allowed a gathering of ten people if they practiced social distancing.

JUNE

They extended the 'stay-at-home' order until June 6. They required travelers to self-quarantine themselves for 14 days if they were traveling from places with high infection rates.

The self-quarantine was ineffective because of a variety of reasons—lack of resources, loss of jobs, loss of revenue, among others.

The COVID-19 infection rate in Hasidic Jews in Orange county was 25% to 28% of the tested population.

New York canceled all the sporting events through the rest of the season and summer.

New York allowed the reopening of essential businesses such as construction, manufacturing, fishing, etc.

By the end of June, New York had not experienced a second spike, such as that which crippled the southern states of Florida, Texas, California, and Arizona.

JULY

As of July 10, the number of cases per day dropped below 1000. As of July 12, New York had performed over 4.4 million tests, with 399,500 confirmed cases and 24,974 deaths. It topped in the US in the number of confirmed cases and deaths. New York City, which had more than 50% of the state's population, had over 50% of the deaths in the state.

Still, New York was recording 500–900 fresh cases daily, despite death rates dropping dramatically.

As of July 12, New York had more than 420,000 cases, with 32,444 deaths, which were more cases than recorded in most Asian countries except for India. The striking challenges faced by New York were similar to what Italy faced in terms of a sudden surge in new cases for which the city was not prepared. COVID-19 overwhelmed the healthcare system and drained the medical professionals. The crisis galvanized the entire country, industry leaders, and the federal government to meet the increasing demands for hospital beds, ventilators, mobile morgues, and medical personnel.

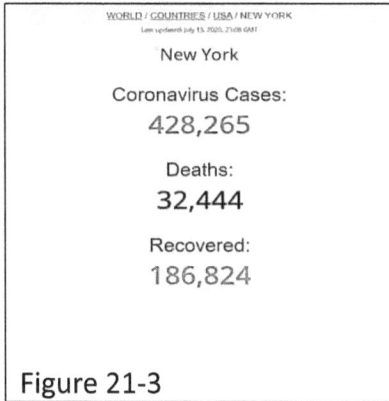

WORLD / COUNTRIES / USA / NEW YORK
Last updated July 15, 2020, 21:08 GMT

New York

Coronavirus Cases:
428,265

Deaths:
32,444

Recovered:
186,824

Figure 21-3

July 15, New York allowed malls to open at 25% capacity.

Never in the history of New York City had anyone seen the Time Square and the 5th Avenue void of people, the subways coming to a screeching halt, and the ballparks left with haunting silence. It looked like a neutron bomb had struck the state!

It was perhaps the biggest challenge New York had faced, a challenge bigger than 9/11, but with resolve and determination, Gov. Cuomo, Mayor DeBlasio, with the Federal assistance from President Trump, had overcome the impossible. They did it their way. "The New York way!"

AUGUST

On Aug 7, New York school opened with certain restrictions.

By Aug 26, the New York infection rate was around 1% of the population tested. It was quite a contrast from the days of mobile morgues outside the hospital, and patients overflowing the hallways.

NEW YORK—IN THE EYE OF THE COVID-19 HURRICANE

COVID-19 was like a hurricane. The city where it makes a landfall gets the maximum impact. New York City sizes and population density also was a factor in the massive devastation. With a hurricane, the airflow in the atmosphere determines where the final landfall is. Here, it was perhaps the movement or flow of people into the city from infected areas. It might be one reason we saw COVID-19's more serious effects in New York City, as it was the receiving port for most of the travelers from the far east.

NEW YORK HOSPITAL CAPACITY

New York had 53,000 hospital beds and 3200 ICU beds at the beginning of the epidemic. Based on some estimates, New York might need 55,000 to 100,000 beds, including 18,600 to 37,200 ICU beds. Some estimated the need for 30,000 more ventilators.

The total number of COVID-19 cases in New York was 37,000 on March 27. Chinese studies had shown that 5% of people might get seriously or critically ill. The Italian data suggested 10% mortality.

Even if you took the worst scenario that 20% of the people getting sick might need ventilators, the number of additional ventilators needed in New York would be 7,400 (20% of 37,000 active COVID-19 cases on March 27).

If the estimates called for 30,000 additional ventilators, then they were expecting to have 300,000 active cases. If you estimated the need for ventilators if 5% of the infected population, then the projections would be for 600,000 active cases, which was not the case in New York in March 2020!

Luckily, they were not there yet. That was the time to get the ventilators ready. If they didn't use them, they could have shipped them to other states that needed them. The media wanted Governor Cuomo to back up his claims with facts and realistic projections.

It would have been useful if New York and other states reported:

1. Total number of daily new cases (done)
2. Total number of daily new admissions to the hospitals
3. Total number of patients on ventilators
4. Total number of new patients needing ventilators
5. Total number of recent daily deaths (done)

Based on these realistic facts

- They could project what was in the current inventory
- How many resources they already had used?
- How quickly the need for additional ventilators was going up?
- How do they meet future needs?

When New York required 16,000 ventilators in 2015, to prepare for a future pandemic such as this one, Gov. Cuomo had refused to act, according to an article in the Western Journal.

Ref:

(https://www.westernjournal.com/cuomo-attacks-trump-exposes-nys-ventilator-shortage-traced-straight-cuomo/)

(https://www.nytimes.com/interactive/2020/03/13/opinion/coronavirus-trump-response.html)

https://nypost.com/2020/03/20/City-hall-didnt-order-covid-19-supplies-for-nyc-until-march-6/

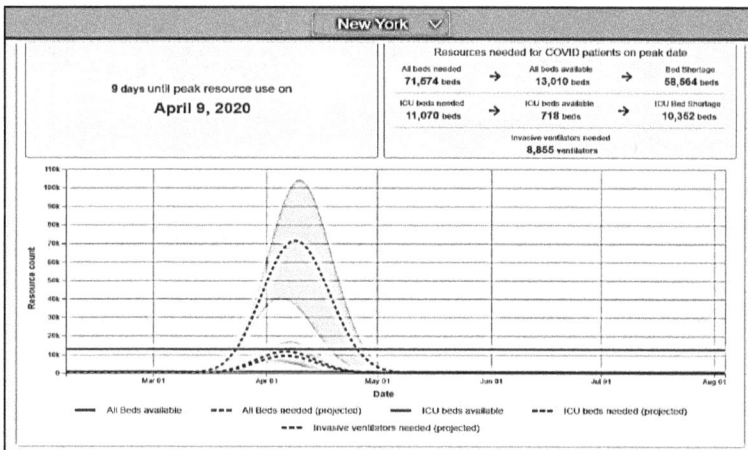

167

COVID-19 PANDEMIC 2019-2020

New York, Catalonia and Madrid could pass Lombardia as the worst affected subnational regions

Cumulative number of deaths, by number of days since 10th death
Showing US states and selected subnational regions in Italy, Spain, China, France, S Korea and UK

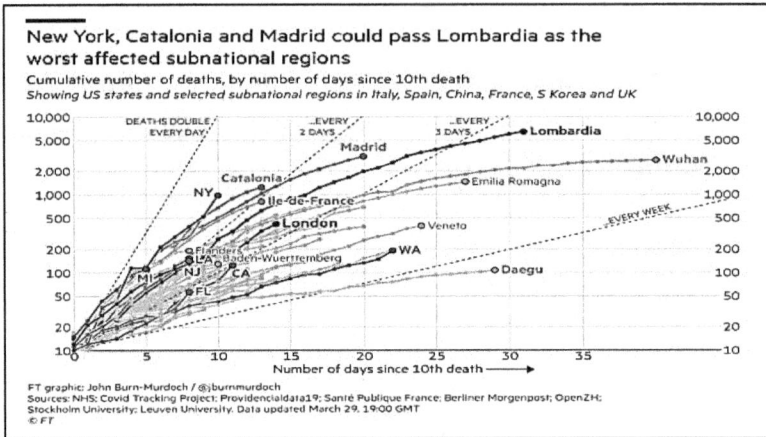

FT graphic: John Burn-Murdoch / @jburnmurdoch
Sources: NHS; Covid Tracking Project; Providencialdata19; Santé Publique France; Berliner Morgenpost; OpenZH;
Stockholm University; Leuven University. Data updated March 29, 19:00 GMT
© FT

Most western countries are on the same coronavirus trajectory. Hong Kong and Singapore have managed to slow the spread

Cumulative number of cases, by number of days since 100th case

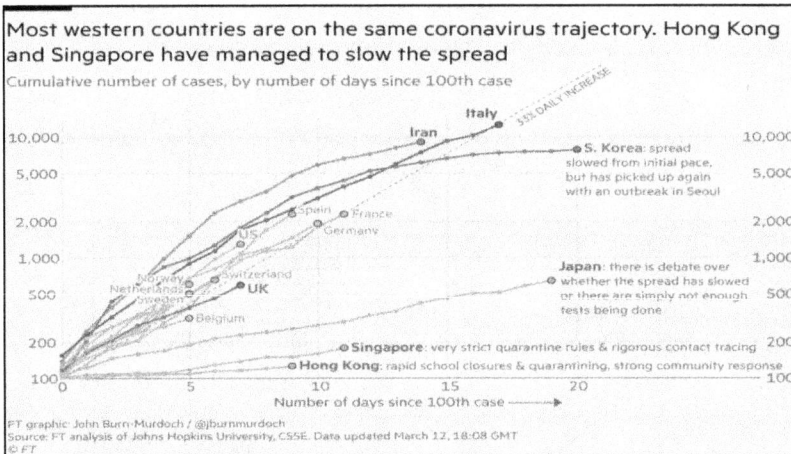

FT graphic: John Burn-Murdoch / @jburnmurdoch
Source: FT analysis of Johns Hopkins University, CSSE. Data updated March 12, 18:08 GMT
© FT

COVID-19 PANDEMIC 2019-2020

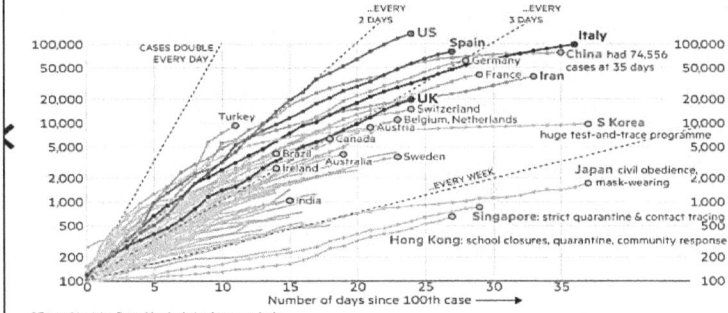

Country by country: how coronavirus case trajectories compare
Cumulative number of confirmed cases,
by number of days since 100th case

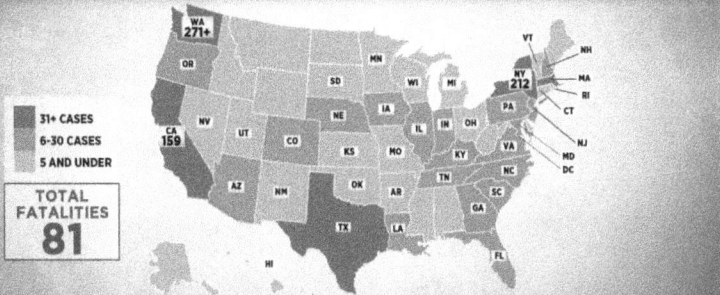

Confirmed Coronavirus Cases in the United States
Total Cases: 4,315

New York City street

NYC ER waiting

Hospitals flooded with patients

Central park tents for patients

Javits center temp hospital

Naval USS Comfort hospital

Ventilator splitting

Never-ending ambulances arrivals

New York City

NY street, no traffic, no people

22 THE UNITED KINGDOM (UK) COVID-19 RESPONSE

The UK, which dominated half the world at one time, had its share of challenges and tribulations in how they handled the COVID-19 pandemic. The UK had a population of 66.65 million. England had a population density of 432 people per square kilometer, while New York City had a population density of more than 44,000 people per square mile. The UK is one of the busiest hubs for international passengers flying from all over the world. (Figure 21-1)

Figure 22-1

On September 12, the UK had 365,174 COVID-19 cases and 41,623 deaths. It had a death rate of 612 per one million, while the US had a death rate of 597 per one million. Even though the total number of cases was less than 10% of the cases reported in the US, their mortality rate was still high compared to many European countries.

The media and public raised questions about how the UK (and many other countries) handled the COVID-19 crisis and if they could have done better. What lessons did they learn from their experiences for future pandemics?

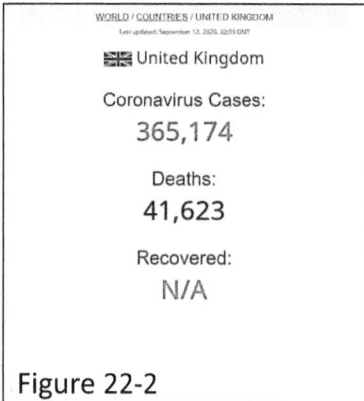

Figure 22-2

WORLD / COUNTRIES / UNITED KINGDOM
Last updated: September 12, 2020, 22:59 GMT

United Kingdom

Coronavirus Cases:
365,174

Deaths:
41,623

Recovered:
N/A

If you do not learn from history and apply those lessons, you are destined to repeat history! Doing the same thing and expecting a different or a novel result is known as the definition of insanity. (Figure 21-2)

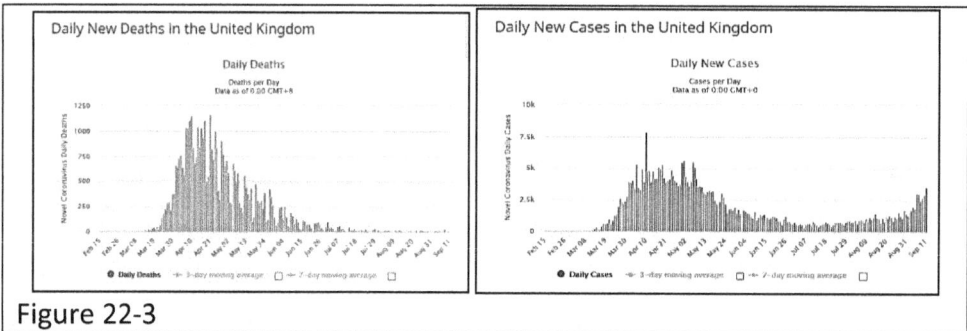

Figure 22-3

THE UK RESPONSE TO COVID-19 PANDEMIC – TIMELINE

The UK reported its first two cases of COVID-19 on Jan 31, 2020. On February 25, the government advised travelers returning from Wuhan, Iran, and South Korea to self-quarantine for two weeks. By March 1, all three British islands had COVID-19 cases (Figure 22-4)

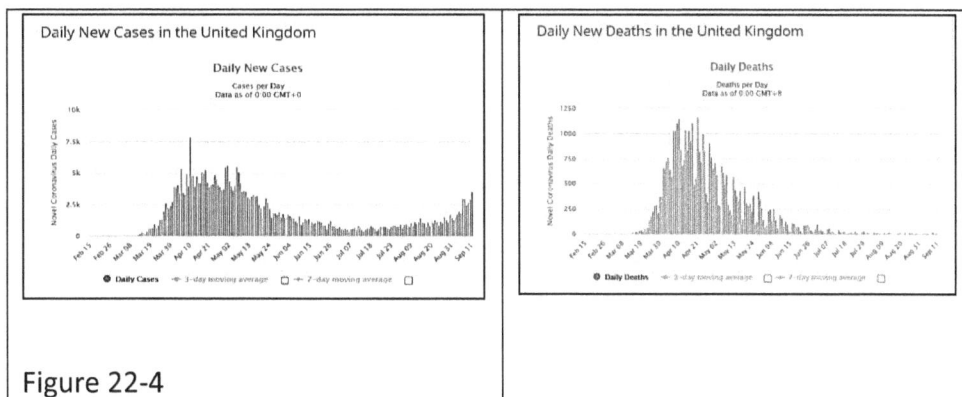

Figure 22-4

The UK reported its first COVID-19 death on March 5, 2020. On March 6, they allocated 46 million pounds for vaccine research and diagnostic tests.

On March 9, the UK stock market index, FTSE, plunged by more than 8 percent. On March 12, it dropped another 10%, its biggest loss since the 1987 stock market crash.

On March 13, the UK suspended all non-urgent surgical and medical procedures, visits, and tests.

The death toll from COVID-19 climbed to 55 with 1500 new cases by March 16, 2020. All individuals over 70 were advised to self-isolate. The government planned for a daily press conference on COVID-19.

The NHS had 8175 ventilators, and it was determined that they might need as many as 30,000. On March 17, the NHS suspended all non-urgent operations to free 30,000 beds to serve COVID-19 patients. They advised against all non-essential travel abroad. The government advised against mass gatherings in sporting events, theaters, etc.

PM Boris Johnson ordered all businesses to close on March 20, including schools and colleges. They canceled all exams for the academic year. New evictions from rental spaces were suspended for three months.

On March 20, Chancellor Rishi Sunak announced that the government would pay 80% of wages to the employees who had lost their jobs.

On March 23, PM Boris Johnson announced a partial lockdown, advising people to stay at home, which went into effect on March 26. All the mobile networks sent an alert instructing citizens to stay at home. By March 25, police were given the power to enforce the lockdown.

On March 27, PM Boris Johnson tested positive for the COVID-19 and entered self-quarantine. On April 5, they admitted him to London hospital for more testing. He had to spend two to three days in the ICU. They discharged him from the hospital on April 12. He applauded the NHS workers for saving his life. He recovered to return home and report to his duty.

Leon retreatants, a fast-food chain, delivered 5,600 free meals to the NHS workers.

On March 30, plans were made to fly tens of thousands of British nationals who had been abroad back home. This was not without a cost, as some were returning from places with high COVID-19 rates.

Studies from Sheffield and Ulster universities showed that anxiety went from 17% to 36%, while the percentage of people who reported depression went from 16% to 38%.

As of April 1, the UK still had a shortage of testing kits. Matt Hancock reported that by the end of April, they should be able to test 100,000 people per day (for a country of more than 60 million people).

On April 3, NHS opened its first ExCel Center and temporary hospital with a capacity of 500 beds that could expand to 4000. Road traffic fell by 73%.

On April 5, Queen Elizabeth II addressed the UK, thanking people for following the government's social distancing guidelines and paying tribute to the NHS workers, assuring them the UK would overcome in its fight against the coronavirus, but that they might have to endure more in the coming days and weeks.

By April 6, the death toll from COVID-19 exceeded 5000 with more than 52,000 active case reports, which put the mortality below 10% of active cases. They revised the need for ventilators to 18,000, but the NHS still had fewer than 10,000 ventilators.

By April 9, UK citizens were reporting signs of cabin fever from the lockdown (2 weeks). On April 10, they recorded the highest number of new cases exceeding 8000.

On April 15, hospital visitation rules were changed to allow close relatives to see their loved ones who were deathly sick. The next day, the lockdown was extended for another three weeks.

As of April 18, the UK was still experiencing PPEs shortages. NHS workers expressed serious concerns about reusing PPE. They were desperately scrambling to get protective devices like masks from many external sources. By this time, the death toll from COVID-19 had reached 15, 464. Care homes in the UK, equivalent to US nursing homes, housed hundreds of thousands of senior residents. They shocked the nation by reporting that 7,500 deaths in care homes had not been accounted for (a similar theme reflected in the US).

The UK announced another 1.6 billion pounds in aid to support local authorities.

They also expressed concerns that the BAME communities (Blacks, Asians, Minorities, and Ethnics) had a much higher number of new cases and deaths than others.

On April 10, they had 18,500 deaths from all causes, 8000 more than average for that time of the year.

On April 20, they locked the front door of buses, eliminated payments, and allowed passengers to board the bus from the middle doors.

On April 22, Prof. Whitty, the government's chief medical adviser (equivalent to Dr. Fauci and Birx in the US) suggested that social distancing would be necessary for at least the rest of the year.

By April 23, the COVID-19 death toll had risen 18,738, and the first human trial on a COVID-19 vaccine began in Oxford. At that stage, the UK was testing more than 50,000 people per day. They were still contemplating instituting contact tracing. They talked about the phased-in lifting of the lockdowns. DIY chain B&Q, similar to Home Depot in the US, reopened.

By April 24, the UK was still working on performing COVID-19 testing on key workers. People flooded the online site to register with more than 5000 requests, and they temporarily shut the site.

On April 25, the UK's COVID-19 deaths passed 20,000.

By April 26, they saw a decrease in the number of new cases and daily deaths. PM Boris Johnson sounded optimistic; they could turn the tides now but urged people to follow self-mitigation guidelines. The government announced a payment of 60,000 pounds for the families of NHS and healthcare providers who died from COVID-19. They recorded total deaths from COVID-19 at 413, which was the lowest in four weeks.

By April 28, they recorded a total death from all causes of 22,351, which was twice the weekly average for the past five years. One-third of those deaths came from care homes.

Since April 10, there has been a steady decline in the number of new cases and deaths, suggesting that the UK was on the downslope of the Bell curve or the Farr curve of the COVID-19 epidemic.

On April 30, PM Boris Johnson said the country was past the peak, but they must try to avoid a second spike.

They reached the capacity to test 100,000 people by the end of April 2020.

On May 3, the NHS introduced a contact tracing app and piloted it in the Isle of Wight. On May 5, they officially opened the NHS Nightingale Hospital Northeast, a temporary critical care hospital. PM Boris Johnson promised the UK would reach a capacity to test 200,000 people by the end of May 2020.

May 7, Scotland extended the lockdown for three more weeks. Wales followed the same. On May 11, PM Boris Johnson released a 50-page document setting the details of the phases for lifting lockdown. On May 13, they eased lockdown measures.

By that time, the UK economy had shrunk by 2%. On May 14, they approved the first COVID-19 antibody test by Roche for use. On May 18, transportations by trains and buses restarted.

COVID-19 PANDEMIC 2019-2020

PM Boris Johnson confirmed that a track and trace system would be in place from June 1. On May 24, the total number of COVID-19 cases in the UK stands at 257,154, with 36,675 deaths.

What lessons did the UK learn that could be applied for future pandemics?

- Most countries were unprepared for pandemics. They need to address that for now and for the future
- When there is a pandemic, start working on the test kits to detect the virus early
- Stockpile PPEs and masks in preparation for pandemics
- Don't downplay the seriousness in the early stages, which only makes things worse, as was the case during the Spanish 'Flu in 1918-1919
- Get the necessary equipment like respirators and ventilators in advance
- Stop pointing fingers. You were all sailing in the same ship of uncertainty. No one was an expert on the new pandemic. First, learn to kill the virus, then talk later
- Self-mitigation is required from the first news release that you have an active case in town
- Lockdown as soon as you reach five or more deaths in the country.
- Stay safe—only you can protect your life
- Continue self-mitigation long after the lockdown, perhaps until a vaccine is available

In the future, the countries should focus on mobile units that can move them from one city to another. Mobile RV hospitals may be efficient and save human labor. These could be mobilized across state lines in a fraction of the time it would take to build a tent hospital.

It is a great lesson for everyone to better prepare for the second wave or the next pandemic.

Lockdown might not be popular when you have as few as 100 new cases or five new deaths, but that is the most vulnerable incubation period. That is the best time to follow self-mitigation and lockdown guidelines to keep the cases and number of deaths down.

Gov Cuomo, on March 18, had said there was no way he was going to lockdown New York, yet four days later, he did. NY got a lot of help from the Federal government compared to that of most other countries. It was a federal and state teamwork that cut the death rate in NY from hundreds of thousands to less than 30,000 at the time of this writing.

As I was collecting data on Germany, and I saw the same politics in Germany in the initial stages. That is a great learning lesson. Countries should develop a checklist to 'smell' the virus from ten thousand miles away and begin action early. (Figure 22-5)

Descriptions	UK	US	Sept 12, 2020
Total cases	365,174	6,670,939	
Deaths	41,623	197,976	
Total cases/1 M pop	5374	20,130	
Deaths/ 1 M pop	612	597	UK higher death rate
Total tests done	19,293,329	91,554,187	
Test/1 M pop	283,905	276,274	UK More tests/ 1M pop
Population	67.85 M	330.1 M	
Population > 65 years	15.7 M	68.7 M	

Figure 22-5

23 HONG KONG SUCCESS STORY—NO LOCKDOWN

Hong Kong is one of the most advanced states in the Southeast part of China. Hong Kong, a former British colony, is now autonomous yet very much controlled by mainland China. In 2020, Hong Kong had a population of 7.5 million, and most people spoke Cantonese. The life expectancy in 2020 was 83 years, with a mean age of 44.4 years. It had a population density of 17,311 people per square mile. (Figure 23-1)

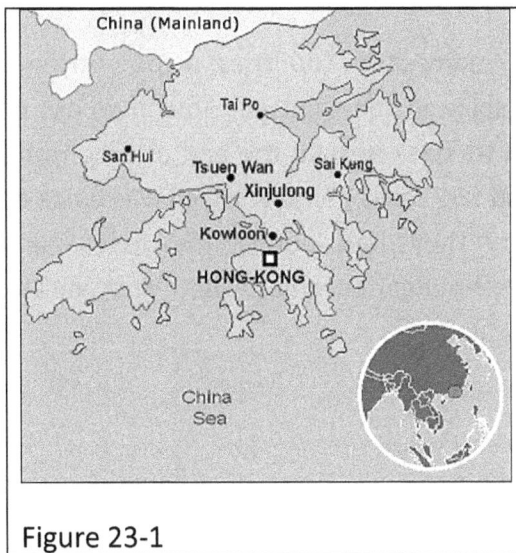

Figure 23-1

Hong Kong has a market-oriented economy and a leader in manufacturing technology-related goods. Many international companies have offices in Hong Kong. Their healthcare system mirrored the UK's National Health Service (NHS) and a single Medical Information System (MIS).

SARS SCARS

Hong Kong was a hot spot during the 2003-2004 SARS epidemic, and the people had firsthand knowledge of what it meant to live and survive during a deadly epidemic. They had 1800 cases with 330 deaths, largely because of the dramatic measures they took to contain the SARS viral spread. The lessons they learned were very fresh in the minds of those who had lived through the epidemic and were well prepared to deal with the COVID-19 epidemic when it hit Hong Kong in the middle of Jan 2020.

COVID-19 EPIDEMIC

Hong Kong reported the first case of COVID-19 on Jan 21. On Sept 12, they had 4926 cases, with 99 deaths related to COVID-19. Their COVID-19 epidemic peaked on March 28, with the daily cases reaching 80 cases. By the end of April, they had a very minimal number of cases of active COVID-19 cases. (Figure 23-2)

WORLD / COUNTRIES / CHINA, HONG KONG SAR
Last updated: September 12, 2020, 09:15 GMT

China, Hong Kong SAR

Coronavirus Cases:
4,926

Deaths:
99

Recovered:
4,597

Figure 23-2

Like many other countries, it too noticed a second spike in the third and fourth weeks of June. At its peak around the end of July, Hong Kong had 150 cases per day, which was twice as many as they had during the first spike. However, with their exceptional self-mitigation rules, they dramatically reduced their cases by the end of August.

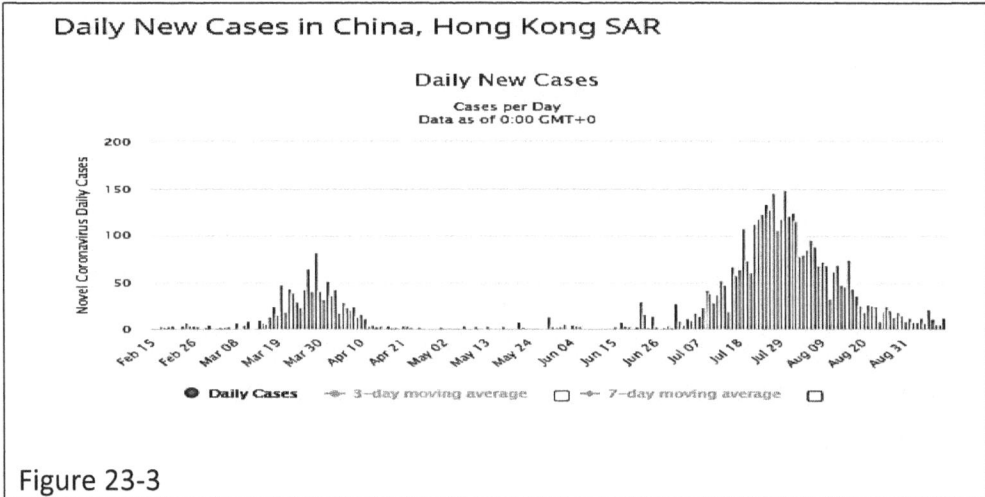

Figure 23-3

Those numbers were nowhere near the 70,000 to 80,000 new cases seen in the US and Brazil during July and August. (Figure 23-3)

LOCKDOWN?

Hong Kong did not have a formal lockdown of the entire state. They close businesses and government offices. However, they had no interruption of local transportation. They restricted air travel to three ports, primarily for the transportation of goods and products. According to a CNBC documentary, more than a million travelers visited Hong Kong during January and February. The restaurants, bars, and clubs were open for business as usual. The schools stayed functional. Local businesses and health clubs were still in operation.

On Aug 25, Hong Kong reported the first case of COVID-19 reinfection.

HOW DID THEY KEEP THE COVID-19 EPIDEMIC UNDER CONTROL?

MASKS. First, most people were wearing masks from the earliest days of COVID-19. They did not even have to institute a law requiring people to wear masks to achieve this. Since those people had lived through the SARS epidemic, it was known in their culture that masks were the most important things to reduce the COVID-19 spread. Hence, masks were everywhere, and every person had one.

CENTRAL COORDINATION OF ALL COVID-19 RELATED MATTERS

Hong Kong had a single central governmental agency (equivalent to the US CDC) that directed and coordinated all COVID-19 related matters. There were no county, city, or regional level regulations controlling at various levels, which made the implementation of its guidelines unified and standardized. COVID-19 matters were never a political issue in Hong Kong, unlike in the US.

BEST QUARANTINE SYSTEM

Hong Kong developed a very rigorous quarantine system for all those coming from outside the country. They tested all incoming passengers. Even if they tested negative, they gave all of them an electronic bracelet that monitored their movements for the next 14 days. They required them to follow the instructions to the letter. If anyone violated the guidelines, they were fined $3200 ($HK) and faced a possible six-month jail sentence. The people in quarantine received regular calls from caseworkers enquiring about their welfare. Additionally, their contacts were traced, and caseworkers followed up with all contacts to mitigate COVID-19 spread.

CROSS IMMUNITY?

Even after protests flared up toward the end of the epidemic in Hong Kong in April, they still had a low incidence of new COVID-19 cases. That raised the question of whether people had some inherent immunity or cross-immunity from being exposed to the SARS virus, which was also a coronavirus. However, the SARS pandemic occurred in 2003, so it is not likely that the population had significant cross-immunity 17 years later.

HEALTHCARE SERVICES

The Hong Kong government-provided healthcare to everyone at virtually no cost. They provided these services not only to citizens but also to non-permanent residents. They had a uniform approach to dealing with the COVID-19 epidemic, with all guidance coming from the head of the Hong Kong CDC.

CAN THE WEST ADOPT SOME OF THEIR PRINCIPLES?

The inherently lower numbers among the East and South Asian countries might have something to do with their geographic locations, exposure to more pathogens, and thus perhaps better immune responses to new challenges. Their genetic makeup and routine use of the BCG vaccination for tuberculosis also could be a factor. These are speculations, and there is no direct evidence for such correlations.

It is worth exploring their quarantine and contact tracing approaches. Hong Kong and New Zealand's approaches to quarantine were some of the best in the world. But, is their strategy applicable to the western world with populations between 50 million and 330 million?

The culture and lifestyle in these two segments of the world were as far apart as the distance that separated them. It would be impractical to impose the level of surveillance used in Hong Kong on people in the western world, where people treasure their rights and freedoms much more so than in the more collective national mentality seen in the far east. In the western world, people protested the lockdowns and the restrictions those placed on their freedom; therefore, it is unlikely that those same people would accept movement tracking.

WHAT ABOUT THE MASKS?

Everyone in Hong Kong wore masks to prevent the spread of the coronavirus. Masks were extremely successful, as reflected in their case numbers and mortality rate. Why can't the West adopt such measures instead of rushing to open the economy, racing to the beaches, and holding COVID-19 parties?

Well, freedom comes with a cost.

We saw such results when Texas experienced the second spike in July and August after lockdowns were lifted and bars, beaches, and clubs opened. The number of younger people with COVID-19 infection admitted to the hospitals rose to 40% of total COVID-19 admissions.

So, not to violate the freedom of other people, the least you could do was to wear masks and reduce the risk of spreading the COVID-19 infection to those people.

We truly cannot ignore the enormous circumstantial evidence for the benefits of wearing masks, as seen in Hong Kong, Japan, Austria, and South Korea.

Ref:

https://www.cnbc.com/video/2020/06/29/how-hong-kong-beat-coronavirus-and-avoided-lockdown.html?&qsearchterm=hong%20kong

https://www.statnews.com/2020/08/24/first-covid-19-reinfection-documented-in-hong-kong-researchers-say/

https://www.nejm.org/doi/full/10.1056/NEJMp2025631

COVID-19 PANDEMIC 2019-2020

WORLD / COUNTRIES / CHINA, HONG KONG SAR

Last updated: June 28, 2020, 15:12 GMT

⚑ China, Hong Kong SAR

Coronavirus Cases:
1,204

Deaths:
7

Recovered:
1,105

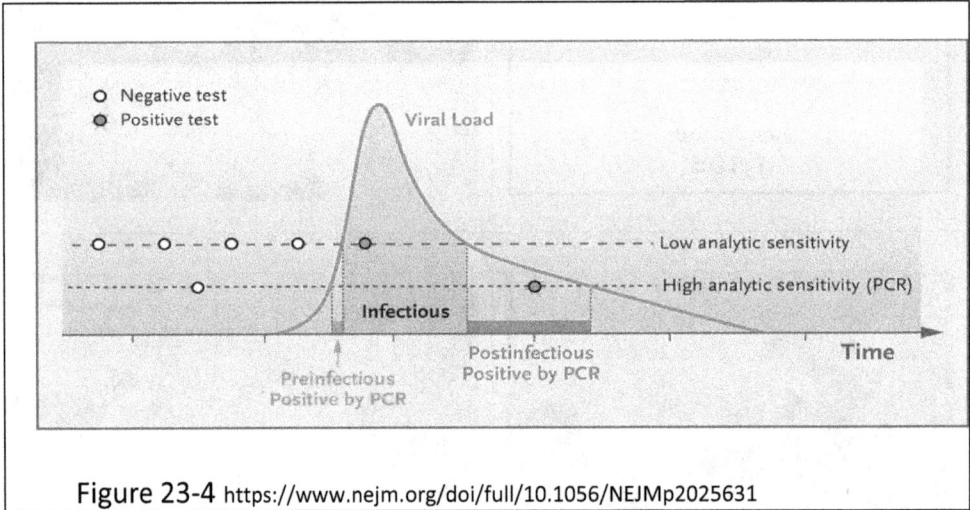

Figure 23-4 https://www.nejm.org/doi/full/10.1056/NEJMp2025631

24 WUHAN AND MAINLAND CHINA

On December 31, the Wuhan CDC announced that a cluster of pneumonia cases with unknown causes had been identified in a Wuhan hospital on December 27. These cases were related to the Huanan Seafood Market and drew nationwide attention, including that of the National Health Commission (NHC) in Beijing. (Figure 24-1)

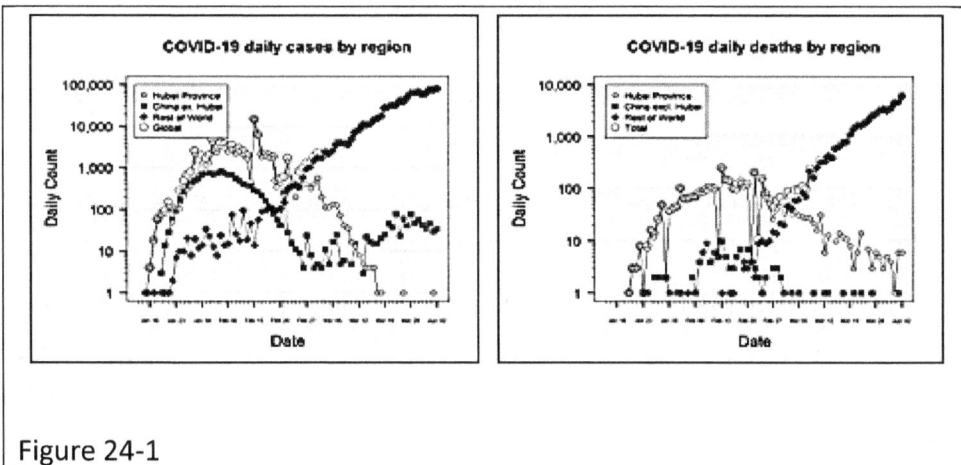

Figure 24-1

China formally notified the World Health Organization on the same day. After the 2003 SARS outbreak experience, China had established an innovative healthcare system to diagnose and manage novel and unknown infections. Within months of the arrival of the H1N1 'flu in 2009, China developed and distributed vaccines to 100 million people.

The pathogen responsible for the 2013 H7N9 outbreak in East China was identified five days after the outbreak, and test kits were designed and distributed to all mainland provinces three days later. Within months, China had an effective vaccine.

JANUARY

On January 2, 41 hospital patients in Wuhan were infected with the novel coronavirus, and 27 of the patients had visited the Huanan Seafood Wholesale Market. These patients were moved to the Wuhan Jinyintan Hospital.

The WHO declared that its China country office, Regional Office for the Western Pacific, and headquarters had been working together to respond to the outbreak.

On January 3, Chinese scientists at the National Institute of Viral Disease Control and Prevention (IVDC) ruled out 26 common respiratory pathogens as the cause of unknown pneumonia. They then identified genetic sequences common to coronaviruses from specimens collected from Wuhan patients and identified the novel SARS-CoV2 virus.

Although not reported to date, health authorities closely monitored the health of 121 close contacts of the initial patients.

On January 3, the Chinese CDC Director, Dr. Gao, discussed initial reports of the virus with the US CDC director Robert Redfield. China formally informed the US Health and Human Services Secretary Alex Azar of the outbreak on March 20.

The head of the University of Hong Kong's Centre for Infection, Ho Pak, informed that it was highly possible that the new illness was spreading from human to human and warned there could be a surge in cases during the upcoming Chinese New Year.

On January 5, the number reached 59 cases, with seven in a critical condition. All patients were quarantined, and their 163 contacts were monitored. At that time, no cases of human-to-human transmission had been observed in healthcare workers.

On January 6, Wuhan health authorities had not ruled out influenza, avian influenza, adenovirus, and coronaviruses SARS and MERS as the cause of the illness. The National Institute of Viral Disease Control and Prevention (IVDC) confirmed it was the novel coronavirus that caused the viral pneumonia of unknown etiology (VPUE) cluster and designated novel coronavirus-infected pneumonia (NCIP).

By this time, the mysterious viral pneumonia was a major topic on social media, and the Chinese government tried to censor those who were spreading misinformation.

The US Centers for Disease Control and Prevention (CDC) created an "incident management system" and issued a travel advisory to travelers to Wuhan, Hubei province, China.

On January 8, scientists published the genetic sequence of the novel coronavirus on an open-access database. South Korea reported its first possible coronavirus case originating from China, and a 36-year-old Chinese woman was isolated amid concerns she had brought the virus from mainland China. On the same date, the WHO confirmed a novel coronavirus had been isolated from one hospitalized patient.

The first coronavirus death, a 61-year-old regular customer of the Huanan Seafood Market, occurred on January 11. He had comorbid conditions, including chronic liver disease, and died from heart failure and pneumonia.

The gene sequencing data of the isolated 2019-nCoV, a virus from the same family as the SARS coronavirus, was shared with GENBANK on January 10. On January 13, the USCDC posted the genome sequence on the NIH genetic sequence database.

On January 12, the WHO published initial guidance on travel advice, laboratory testing, and medical investigation. It echoed the Chinese government reports there was not yet any clear evidence that the virus could be easily transmitted from person to person. On January 14, the WHO tweeted: "Preliminary investigations conducted by the Chinese authorities have found no clear evidence of human-to-human transmission of the novel coronavirus (2019-nCoV) identified in Wuhan, China."

The first novel coronavirus case in the US was identified on January 15 in a woman from Washington state who had recently returned from Wuhan. Her diagnosis was confirmed with a real-time reverse transcription-polymerase chain reaction (rRT-PCR) test for the virus.

The US Embassy in China issued a Level 1 Health Alert Watch for the novel coronavirus.

On January 16, Japan alerted the WHO on a novel coronavirus case in a 30-year-old Chinese national who had returned from Wuhan.

The German Center for Infection Research (DZIF) at Charité – Universitätsmedizin, Berlin, quickly developed a new laboratory test to detect the novel virus.

On January 17, Thailand reported its second case, a 74-year-old woman who had recently returned from Wuhan.

The USCDC dispatched 100 people to three American airports to screen travelers coming from Wuhan, China.

On January 18, Wuhan held an annual banquet in the Baibuting community celebrating the Chinese New Year, attended by forty thousand families, and by January 19, the total number of cases had risen to 201, including one in Guangdong and two in Beijing.

On January 20, Xi Jinping made his first public remark on the outbreak and spoke of the need for the timely release of information. China's National Health Commission announced confirmation that the coronavirus was transmitted between humans. This was confirmed after two medical staff were infected in Guangdong. More cases were identified outside of Wuhan in

Hong Kong, Singapore, the US, and Malaysia, confirming human to human transmission.

After 300 confirmed diagnoses and six deaths, the Chinese state media warned officials not to cover up the spread of the new coronavirus spread. They warned that "anyone concealing new cases would be nailed on the pillar of shame for eternity."

Multiple cities within China reported cases of pneumonia caused by the new virus. A quarantine of Wuhan began on January 23. All public transportation, including buses, metro, ferry lines, and all outbound trains and flights were suspended. The mayor of Wuhan admitted that over 5 million people traveled out of the city before the quarantine.

On January 25, Xi Jinping described the "accelerating spread" of the coronavirus as a "grave situation" and expressed concern that the virus was "mutating." Beijing escalated measures to contain the illness. Chinese canceled New Year celebrations were in many cities.

January 26, China extended the Spring Festival holiday to contain the coronavirus outbreak, Beijing halted all inter-provincial bus and train services, and the USCDC raised the travel advisory for China to Level 3

On January 27, the lethality of the virus was still unknown; however, the death toll had already climbed above three percent. The WHO Director-General Tedros Adhanom Ghebreyesus visited China to discuss the outbreak with senior Chinese officials.

On January 28, China's Supreme People's Court vindicated Dr. Li Wenliang, who had been censored by the Chinese government after reporting the existence of unexplained pneumonia in the early stages of the COVID-19 outbreak, and the entire Hubei province was quarantined, except for the Shennongjia Forestry District.

On January 30, the WHO declared the virus an International Public Health Emergency of International Concern" (PHEIC), reversing two previous decisions after emergency committee meetings. All countries were advised to be prepared for containment, active surveillance, early detection, isolation, case management, contact tracing, and prevention of COVID-19 infection.

India confirmed its first case of coronavirus in a student who had returned from Wuhan University to the Indian state of Kerala.

On January 31, the UK and Russia confirmed their first coronavirus infections. The US banned the entry of foreign nationals who had been to China in the previous 14 days. Italian Prime Minister Giuseppe Conte stated in a press conference that Italy had closed all air traffic to and from China, and Italy became the first E.U. country to a state of emergency due to COVID-19.

In late January, economists had predicted a V-shaped recovery, which was far from a reality in March. There was a sharp drop in many manufacturing indices due to lockdown. The economist predicted a 1.1% drop in China's economy in the first half of 2020. In reality, in the first quarter of 2020, China's GDP dropped by 6.8%.

FEBRUARY

As late as mid-February, the WHO Emergency Committee declared that the virus was not an official Public Health Emergency of International Concern, but warned it was an emergency in China.

The sale of new cars in China dropped by 92% in the first two weeks of February 2020.

In January and February 2020, during the peak of the epidemic in China, approximately 5 million Chinese people in China lost their jobs, and nearly 300 million rural migrant workers were stranded at home.

Personal protective equipment was strongly recommended for healthcare workers dealing with patients. China had a huge shortage of face masks and other protective gear, despite being the world's manufacturing hub for these products. It became clear that testing and careful surveillance for COVID-19 cases was necessary due to its potential to become a pandemic. The virus had reached Australia, Canada, Sri lank, Europe, and many other countries. The WHO said, "The time is now to 'act as one'" in fighting the virus."

China instituted nationwide screening, identification, and immediate isolation of coronavirus-infected travelers at airports, railway stations, bus stations, and ports.

Air Canada halted all direct flights to China until February 29. The Trump administration trade advisor Peter Navarro warned that coronavirus could "Develop into a full-blown pandemic, endangering the lives of millions of Americans" and stated that the "risk of a worst-case pandemic scenario should not be overlooked."

By February 8, worldwide infection rates had exceeded 34,878, with 724 deaths.

On February 25, the number of newly confirmed cases outside mainland China exceeded those from within.

MARCH

By March 6, in China, the number of new cases had dropped to fewer than 100 per day, down from thousands per day at the height of the crisis. However, cases were rising sharply in other countries such as Italy. On March 13, China sent medical supplies, including masks and respirators, to Italy, together with a team of Chinese medical staff. Chinese billionaire and Alibaba co-founder Jack Ma also donated 500,000 masks and other medical supplies to Europe.

By the end of March, 80 million workers were unemployed.

APRIL

On April 2, 2020, the government ordered a Hubei-like lockdown in Jia County, Henan, after a woman tested positive for the coronavirus.

On April 9, Heilongjiang Province reported a cluster of cases, which started with an asymptomatic patient returning from the US and quarantining at home.

MAY

As of May 12, there were 1692 (64 active cases) imported cases in mainland China, with no recent deaths.

JUNE

On June 11, officials confirmed another outbreak of the infection in Fengtai, a south-western district in Beijing.

Later in June, another outbreak involving 45 people at Xinfadi Market in Beijing raised an alarm. Authorities closed the market and nearby schools and instituted temperature checks.

JULY

On July 26, China saw 61 new cases, its highest number of daily cases since March, which increased to 127 cases on July 30, before it dropped to 16 cases on August 23.

Multiple countries evacuated their citizens from Wuhan, including South Korea, Japan, the US, the UK, Kazakhstan, Germany, Spain, Canada, Russia, the Netherlands, Australia, New Zealand, Indonesia, France, Switzerland, and Thailand.

Officials in Spain, Turkey, and the Netherlands had rejected Chinese-made equipment as defective.

The Dutch Health Ministry recalled 600,000 face masks made in China. The Spanish government said almost 60,000 test machines did not produce accurate results. The UK paid two companies in China $20 million for test kits later found to be faulty.

Hotels and restaurants turned away Wuhan natives.

Chinese people overseas experienced discrimination and anti-Chinese sentiments during the coronavirus outbreak. Many restaurants in neighboring countries shunned Chinese customers and refused to serve them. Discriminatory memes about Asian people eating bats went viral on the internet.

Similarly, Chinese people targeted African people on the false premise that they were spreading the infection. People outside China assumed that China had grossly under-reported the number of cases and deaths, even though no primary knowledge was available as coronavirus research was, presumably, censored in China. Theories circulated that the coronavirus originated from bats in the wet market or that it was leaked from a nearby virology research laboratory. The younger generation in China discounted official media reports as propaganda and created mistrust.

The World Health Organization called the Wuhan lockdown unprecedented and said it showed the commitment of authorities to containing viral outbreaks. However, the WHO did not recommend that the US follow a Wuhan-style lockdown approach. After northern Italy became a hotspot in late February, the Italian government did enact such a lockdown of the Lombardy and Veneto regions. Soon after, Iran instated a lockdown of Tehran, with a population of 8 million quarantined.

DR. LI WENLIANG

Dr. Li Wenliang, a Wuhan ophthalmologist, was summoned to the Wuhan Public Security Bureau, where he was told to sign an official confession letter promising to cease spreading false "rumors" regarding the coronavirus.

They reproached him for "making false comments by announcing the confirmation of 7 cases of SARS at the Huanan Seafood Wholesale Market," which had "severely disturbed the social order."

The letter stated, "We solemnly warn you: If you keep being stubborn with such impertinence, and continue this illegal activity, you will be brought to justice—is that understood?"

Li signed the confession writing: "Yes, I understand."

Dr. Li Wenliang developed a dry cough on January 10. He entered the hospital with a fever on January 14. Even though he repeatedly tested negative, he eventually tested positive for the coronavirus on January 30. He died on February 7.

After Li Wenliang died COVID-19 on February 7, 2020, people mourned his death and demanded the Wuhan government issue an apology to him. Later in March, Wuhan police apologized to Li Wenliang's family and praised the whistleblower's effort on raising public awareness.

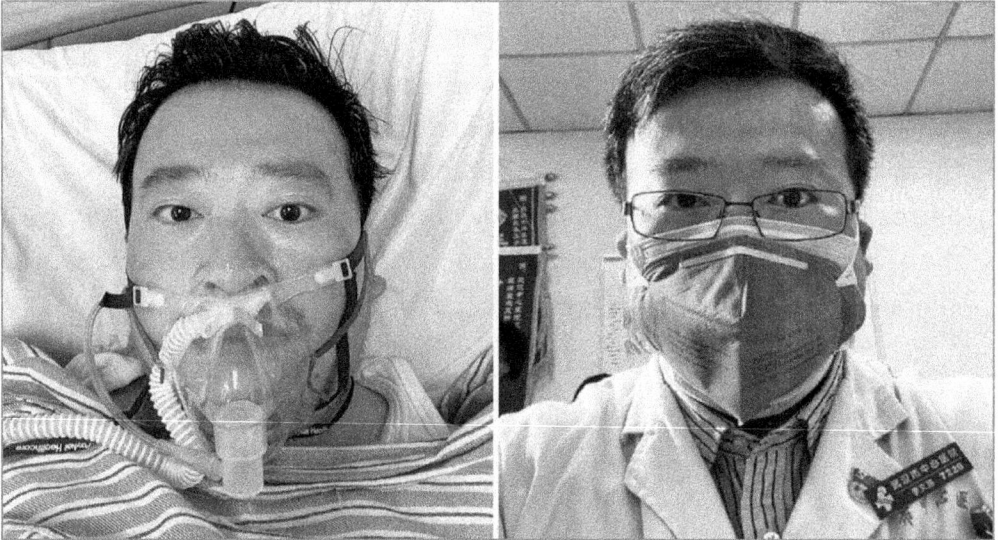

Figure 24-2

25 UNITED STATES' RESPONSE TO FOR COVID-19 JAN TO OCT

The US recorded its earliest cases in the middle of January 2020 in Washington state. New York City became the epicenter of the COVID-19 pandemic in the US. As of Oct 22, 2020, over 8 million people in the US had contracted COVID-19, with more than 220,000 deaths—the highest numbers in the world. But don't let those numbers deceive you.

I will explore in detail the logical reasons for these high numbers and how they compare to European counterparts, as 63% of the US population is of European descent.

Even though we have become accustomed to seven-second sound bites in the media, much of the hype fails to address the fixed variables and risk factors for COVID-19 that are unique to the US and how these contributed to the incidence and mortality rate.

Here are some variables:

- Ethnicity: European, Hispanic, African American, etc.
- Number of TESTS performed
- Age: People over 60 years

- Risk factors: Diabetes Mellitus, hypertension, and cardiovascular disease
- Population density
- Self-mitigation regard | disregard
- Lockdowns and post-lockdowns
- Cultural differences: PEOPLE'S RIGHTS First Amendment
- Strikes, large gatherings, protests,

ETHNICITY

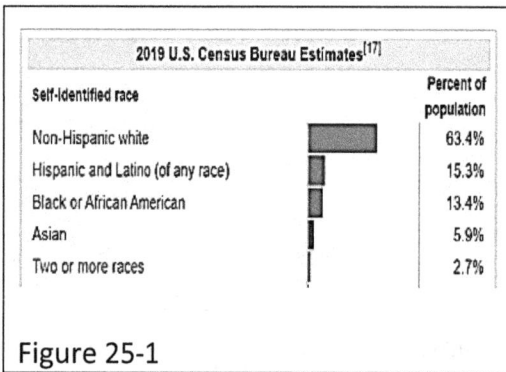

2019 U.S. Census Bureau Estimates[17]	
Self-identified race	Percent of population
Non-Hispanic white	63.4%
Hispanic and Latino (of any race)	15.3%
Black or African American	13.4%
Asian	5.9%
Two or more races	2.7%

Figure 25-1

Over 63% of the US population is of European descent. Hence, we should compare outcomes to those from the UK and other European countries such as France, Italy, Germany, Poland, Belgium, etc. (Figure 25-1)

POPULATION SIZE NUMBER OF TESTS AND

The US, with a population of 331 million people, is the third-largest country after China and India. At the time of writing, the US had tested 39% of the population. More testing means that more of the positive cases are identified. The US may not have the highest incidence of COVID-19 cases, as the media might like you to believe. The 8.5 million positive cases represent just 2.6% of the population, meaning that 97.4% of the population was uninfected. (Figure 25-2)

WORLD / COUNTRIES / UNITED STATES
Last updated: October 21, 2020, 91:34 GMT

United States

Coronavirus Cases:
8,520,307

Deaths:
226,149

Recovered:
5,545,619

Figure 25-2

PEOPLE OVER AGE 60 YEARS

The incidence of COVID-19 is extremely low in people under 29 years of age, with mortality around 0.3%. As age increases, so does the risk of COVID-19 infection and death. The hospital admission rates are 4 -12 times higher in the 60-85+ age groups. (Figure 25-3)

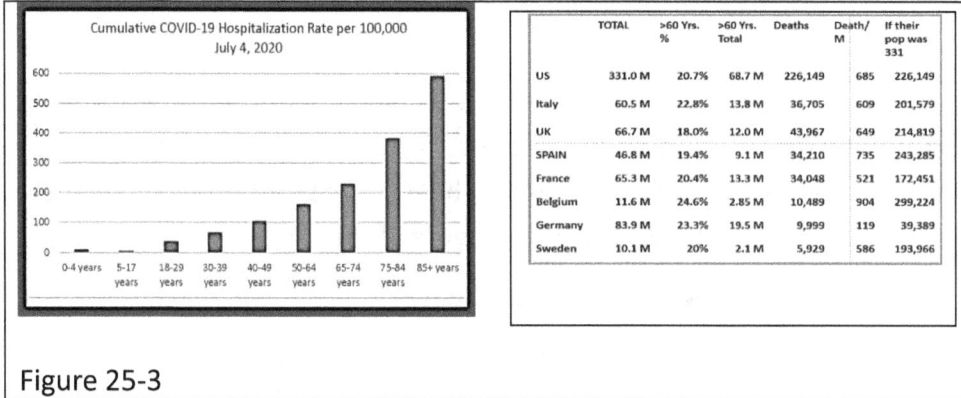

	TOTAL	>60 Yrs. %	>60 Yrs. Total	Deaths	Death/ M	If their pop was 331
US	331.0 M	20.7%	68.7 M	226,149	685	226,149
Italy	60.5 M	22.8%	13.8 M	36,705	609	201,579
UK	66.7 M	18.0%	12.0 M	43,967	649	214,819
SPAIN	46.8 M	19.4%	9.1 M	34,210	735	243,285
France	65.3 M	20.4%	13.3 M	34,048	521	172,451
Belgium	11.6 M	24.6%	2.85 M	10,489	904	299,224
Germany	83.9 M	23.3%	19.5 M	9,999	119	39,389
Sweden	10.1 M	20%	2.1 M	5,929	586	193,966

Figure 25-3

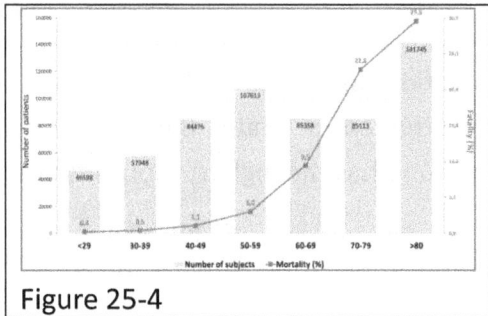

Figure 25-4

Similarly, mortality in people aged over 60 is 32-100 times that of people under 29. Figure 25-4 shows that 68.7 million people in the US are over 60 years of age, which is greater than the population of many European countries. That explains the high incidence and mortality among those over the age of 60.

DIABETES, HYPERTENSION, AND HEART DISEASE

Diabetes mellitus, heart disease, and hypertension were identified as major risk factors in the Italian group and are associated with extremely high mortality

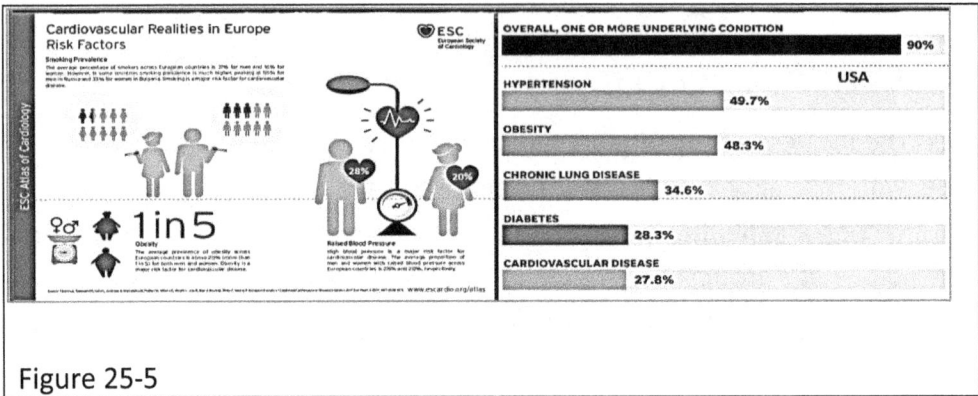

Figure 25-5

The mortality rate in people aged 65+ who had all three of these risk factors was 50%-75%. These are fixed risk factors. Only 20% of the European group had these risk factors. In the US, heart disease, obesity, and diabetes prevalence ranged from 27% to 50%. (Figure 25-5)

POPULATION DENSITY AND MORTALITY

There is a direct correlation between population density and COVID-19 mortality. Mumbai has a population density of 73,000 per square mile and total mortality of 42,831, with a population of only 18.41 million. Despite India having low mortality, the city of Mumbai is an exception. You cannot change the population density.

Similarly, New York City has a population density of 26,403 per square mile and total mortality of 33,523 for the entire state. Belgium has a population density of 991 per square mile and has the highest mortality rate in the EU block. (Figure 25-6)

POPULATION DENSITY PER SQUARE MILE				
Names	DEATH/ M	Density/Sq ml	MORTALITY	POP
NYC	1723/ M	26,403	33,523	19.5 M
Italy	609/ M	532	36,832	60.4 M
UK	649/M	432	44,158	67.99 M
France	521/M	309	34,048	65.32 M
Germany	119/M	623	9,999	83.9 M
Belgium	904/ M	991	10,489	11.6 M
Sweden	586/ M	64	5,929	11.6 M
Mumbai	2326/M	73,000	42,831	18.41 M

Figure 25-6

LOCKDOWNS AND SELF-MITIGATION

We saw positive results from lockdown and self-mitigation measures in countries such as Hong Kong, Japan, and Germany. Lockdowns were also successful in the US initially. However, the mass disregard for self-mitigation, major gatherings, strikes, and protests were beyond state or federal control and contributed to COVID-19 spread.

Americans view their freedom of speech and expression passionately, which perhaps partially contributed to the increased incidence. In Texas, we saw a sudden surge in COVID-19 cases, hospital admission, and even younger people getting the infection after the lockdown was lifted.

Human beings are very independent, and their feelings and perceptions can swing from one extreme to the other. Despite lockdowns and self-mitigation measures, we are also seeing a second wave in other parts of the world, including the UK and EU counties.

This recent surge of cases all over the globe with declining mortality has been termed "casedemic," where the PCR tests for COVID-19 are detecting viral RNA instead of live virus particles, leading to cases being identified that do not represent active infections capable of spreading the disease. A test that only detects the live virus may provide a more accurate picture of active infection rates.

Unlike many EU countries and the UK, where healthcare is nationwide and centralized, healthcare in the US is fragmented, controlled by private entities, and under state regulations. Hence, the approach varied from state to state with a touch of political influence base on party affiliations.

In conclusion, the US has the third-highest population in the world, it has the highest population over the age of 60 years compared with the EU countries, it has a much higher rate of obesity, diabetes, heart disease, and hypertension, and has a highly fragmented healthcare system. It is also a country where people are passionate about their freedom of speech and expression, including the right to protest.

Despite all these challenges, the US has traveled light-years in the development of new therapeutics, antibodies, hyperimmune globulins, plasma transfusion, and over half a million ventilators. There are several vaccines in late Phase-3 trials for the SARS-CoV-2 virus.

In January and February, the initial projections of mortality in the US were close to a million, and that projected number has decreased dramatically over the course of the pandemic.

Still, a lot needs to be done in terms of contact tracing and containing this virus, including eliminating the threat with a vaccine in the coming months. Given the limitations, the US has responded well, and the outcome is comparable to its European counterparts.

ACTIONS TAKEN BY PRES. TRUMP'S ADMINISTRATION TO FIGHT CORONAVIRUS.

ENTIRE GOVERNMENT RESPONSE:

In January, the coronavirus was declared a public health emergency. A national emergency was declared and $8.3 billion was secured for the coronavirus response.

The Families First Coronavirus Response act was put in place to help American families and businesses impacted by COVID-19 receive the support they needed.

The White House Coronavirus Task Force was led by Vice President Mike Pence, along with Dr. Deborah Birx. President Trump held many teleconferences with the nation's governors to coordinate response efforts and to assure federal support. Pres. Trump approved major disaster declarations for severely impacted states, such as New York, Washington, and California.

TRAVEL RESTRICTIONS

In January, President Trump issued a travel ban from China. Further travel restrictions from the European Union, Ireland, and Iran were instituted in February. American citizens returning from abroad were directed to specific airports to be screened and isolated if necessary.

Non-essential travel across northern and southern borders was restricted. The highest-level travel warnings were issued to hot spots like Japan and South Korea. A global level 4 travel advisory urged people to avoid all international travel.

TESTING AND REPORTING

The Food and Drug Administration issued emergency approval for new commercial coronavirus tests. Health and Human Services provided funds to accelerate the development of rapid diagnostic tests for COVID-19. The Department of Defense set up 15 coronavirus testing sites worldwide.

The president signed legislation requiring reporting from state and private labs to ensure accurate data gathering to respond to the outbreak. DoD and HHS worked to airlift hundreds of thousands of swabs and sample test kits from Italy to the US. The US-manufactured more than 500,000 ventilators.

SUPPORTING BUSINESSES

President Trump directed the Energy Department to purchase significant quantities of crude oil to supplement the strategic oil reserve. He held meetings with business leaders from across the board, including pharmaceutical, airlines, health insurers, grocery stores, retail stores, banks, etc. The Treasury Department approved the establishment of the Money Market Mutual Fund Liquidity Facility to provide liquidity to the financial system.

HELPING FAMILIES AND WORKING AMERICANS

Provided tax credits for eligible businesses providing paid leave to Americans affected by the virus. Provided funding and flexibility for emergency national aid for senior citizens, women, children, and low-income families.

The USDA announced a collaboration with the private sector to deliver nearly 1,000,000 meals per week to students in rural schools. Foreclosures and evictions for families with FHA-insured mortgages were halted.

A website was launched for families, students, and educators to access online technology. Legislation was put in place to provide continuity in education for veterans and their families to switch from in-person learning due to the coronavirus. The Department of Education gave approval to colleges and universities to move classes online. The Department of Education suspended payments on federal student loans until the end of the year, still interest-free. Federal testing requirements were waived due to school closures.

THE PUBLIC EDUCATION

The president launched a partnership with the Ad Council, media networks, and digital platforms to communicate public service announcements about the coronavirus. Guidelines were published for Americans to follow to stem the spread of the virus.

CMS announced guidance to protect vulnerable elderly Americans and limit medically unnecessary visits to nursing homes. The president authorized the HHS to waive rules and regulations, so the healthcare providers had maximum flexibility in responding to the outbreak.

The White House Office of Science and Technology Policy coordinated with the NIH, the tech industry, and nonprofits to release a machine-readable collection of 29,000 coronavirus research articles to help scientists discover insights into virus' genetics, incubation, treatment symptoms, and prevention.

The Veterans Administration established 19 emergency operations centers and visitation restrictions to limit patients' exposure. The VA limited nonessential, elective medical procedures. Two Navy hospital ships, the Mercy and the Comfort, were moved to New York and Los Angeles to add additional beds. Carnival Cruise Lines lends their ships for use for non-COVID-19 patients.

STRENGTHENING ESSENTIAL MEDICAL SUPPLIES

President Trump invoked the Defense Production Act. The president signed a memorandum to make general-use face masks available to healthcare workers. HHS announced it would purchase 500 million N95 respirators for the Strategic National Stockpile. President Trump signed legislation requiring manufacturers to sell industrial masks only to hospitals and medical personnel.

DEVELOPMENT OF VACCINES AND THERAPEUTICS

The administration worked to accelerate the development of therapeutics and a vaccine, working alongside drug manufacturing companies to ensure the drug supply chain remained intact.

Ref:

BY LEISHA S. D'ANGELO ON OCTOBER 26, 2020

Nik Nikam

June 25 at 9:17 PM · transmission

FACE MASKS, AND MASK MYTHS UNMASKED!
NIK NIKAM, MD, MHA. HOUSTON, TX
COVID-19 UPDATE

There has been plenty of debate about wearing masks to prevent the spread of coronavirus. At the start of this pandemic, we didn't really have much scientific evidence. And what little evidence we had involved other diseases like the flu, SARS, MERS, which can't be applied to the COVID-19 virus.

The first purpose is to prevent a person with coronavirus infection from spreading the virus to other ...
See More

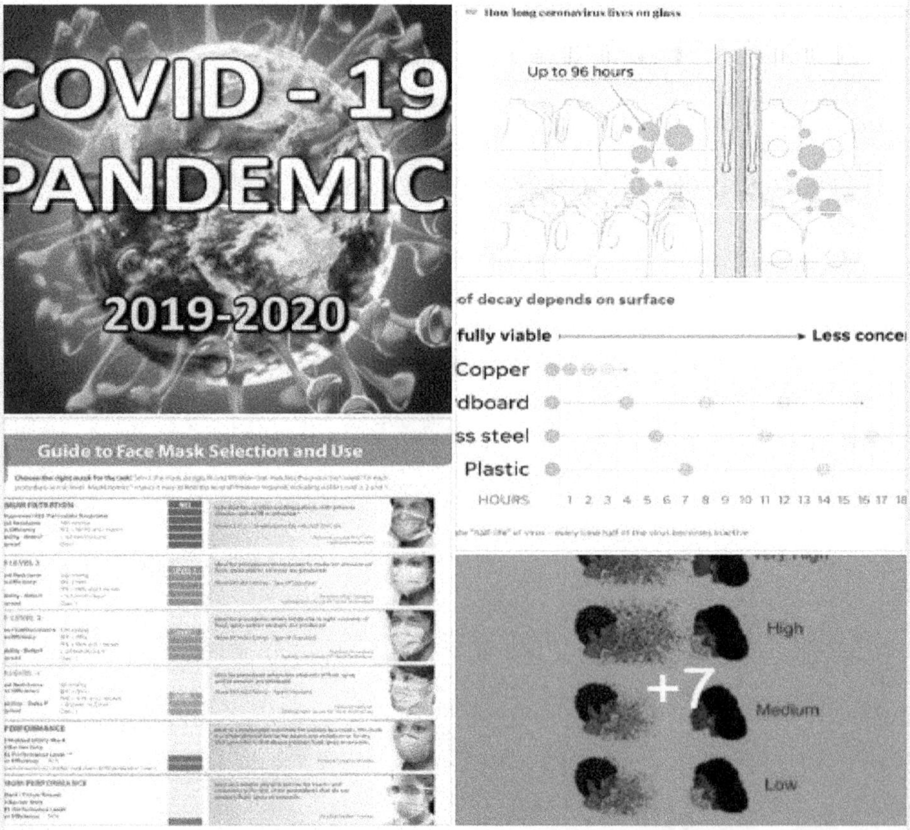

Priya Saini Verma and 248 others

67 Comments

26 GERMANY'S SUCCESS STORY

Germany is a Western European country with over two millennia of

Figure 26-1

history. Berlin, its capital, is home to art and nightlife scenes, the Brandenburg Gate, and many sites related to WWII. Munich is known for its Oktoberfest and beer halls, including the 16th-century Hofbräuhaus. Frankfurt, with its skyscrapers, houses the European Central Bank. It has a population of 83 million people, with 23 million over the age of 60 years. (Figure 26-1)

Germany had the best response to the COVID-19 pandemic in terms of identifying the new active cases, isolating them, performing contact tracing, and public education. They served as a model not only for EU countries but for the rest of the world. However, despite their technologically advanced, pro-active approach, and substantial support from their people and the government, Germany was not immune from the second spike that rocked EU countries in

WORLD / COUNTRIES / GERMANY
Last updated September 19, 2020, 17:15 GMT
Germany
Coronavirus Cases:
271,840
Deaths:
9,466
Recovered:
243,000

Figure 26-2

August and September. (Figure 25-6)

Germany recorded its first COVID-19 case on Jan 27, near Munich, Bavaria. The patient had contracted the virus from a Chinese colleague who had visited Shanghai. Most cases in January and early February came from the same automobile-parts plant.

As early as January, experts informed the German Bundestag of the magnitude of the COVID-19 pandemic and high mortality rate in individuals aged over 65. The Federal Office of the Civil Protection and Disaster Assistance failed to take appropriate steps to prepare. At that time, the German government considered COVID-19 to be a "very low health risk" less dangerous than SARS and did not implement travel restrictions.

On January 28, Jens Spahn, the Federal Minister of Health, said he was more worried about conspiracy theories than the virus and indicated that the federal government would be fully transparent. After identifying a suspected case in a Lufthansa plane, the company suspended all flights to China.

On February 26, following confirmation of multiple cases, Heinsberg closed schools, libraries, and gaming places. They still did not institute travel restrictions to Italy, the epicenter of COVID-19 in Europe.

Germany reached the Top-Ten List of total case numbers in Europe by the end of February, second to Italy.

On March 1, the number of confirmed cases in Germany almost doubled each day. On March 2, the Robert Koch Institute (RKI) raised the threat level to moderate, while the European Centre for Disease Control (ECDC) raised the level from moderate to high.

Berlin reported its first COVID-19 case on March 2. It announced the opening of a 1000 bed hospital dedicated to COVID-19 patients, which became operational on May 11.

As of March 3, healthcare workers reported a lack of protective equipment to handle COVID-19 patients. In response, on March 4, Germany banned the shipment of PPE outside of Germany.

On March 6, German health minister Spahn declared travel restrictions within the EU unnecessary, and at the time, the EU and RKI believed that healthy individuals did not need to use masks and disinfectants.

A 14-day quarantine for those who had been in contact with COVID-19 patients was imposed in Germany on March 6. People traveling from China, South Korea, Japan, Italy, and Iran were required to report their health status before entry.

A large cluster of infections was linked to a carnival in Heinsberg, and Germany reported its first COVID-19 death on March 9, by which time the total case numbers exceeded 1200.

Until that time, Chancellor Angela Merkel had stayed behind the scenes. Facing political pressure, she announced a press conference on March 11 and announced liquidity support to companies. She was not keen on closing the borders. Again, the government declared that mouth protection and disinfectants were needless for healthy individuals and indicated that handwashing with soap should be adequate.

Chancellor Merkel announced 750 billion Euros for vaccine research on May 6. On May 13, border control with several neighboring countries was relaxed.

The US President Donald Trump introduced a travel ban from the Schengen area and Germany on March 12, excluding the UK.

Germany ordered 10,000 ventilators from Dragerwerk on March 14, which was equivalent to typical ventilator orders for an entire year. By that time, Germany had 4,585 COVID-19 cases and nine deaths.

The German economy shrank by 2.2% in the first quarter of 2020. Still, Germany began the Bundesliga football league on May 16.

Germany felt well prepared and did not initially take special measures to stockpile medical supplies or limit movement. By March 13, the Robert Koch Institute (RKI) mandated school closures and prohibited nursing home visits--an important lesson for all countries to make sure the elderly are safeguarded early in an epidemic.

Germany closed borders to five neighboring countries on March 15. A week later, curfews were implemented in six German states. In the other six areas, contact with more than ten people was prohibited. By March 23, Germany's total coronavirus cases had climbed to 179,986, with 8,366 deaths.

Germany canceled all flights from China and Iran on March 16. On the same day, Chancellor Merkel announced a severe restriction of people's movements and business operations but stopped short of implementing a full lockdown. The construction of a 1000 bed COVID-19 hospital near Bundeswehr was announced along with plans for doubling the country's ICU capacity by 28,000, including 25,000 ventilators.

The COVID-19 spread to Germany from other regions. People from China, Iran, and Italy were arriving in Germany until March 18. By March 20, more than 80% of private doctors in Germany reported having trouble procuring PPE.

Germany attempted to manage the early spread in the containment stage by minimizing the expansion of clusters. Chancellor Angela Merkel went into quarantine on March 21 when her physician tested positive for COVID-19. Germany took on a 750 billion euros debt to mitigate the damages from the COVID-19 pandemic on March 23. Masks or face coverings were finally mandated on March 31, although the country was short of these supplies.

LOWER DEATH RATES?

Experts attributed lower mortality in Germany compared with other EU countries to:

- ✓ A higher number of tests done and contact identification
- ✓ A higher number of ICU beds
- ✓ More respiratory and ventilatory supports
- ✓ Absence of COVID-19 analyses in autopsies
- ✓ A higher number of younger people testing positive
- ✓ The way the primary cause of death was reported

The German government established a central PPE acquisition scheme to create a reserve. This is an important lesson for central or federal governments. It is vital to stockpile enough PPE for the entire country in preparation for an epidemic or pandemic.

On April 1, Germany adopted a contact tracing app used in eight EU countries to track and trace COVID-19 exposed individuals and inform them of their need to quarantine.

On April 16, after consulting with 16 Minister Presidents of the 16 Federal states, Chancellor Merkel said that Germany had achieved "fragile intermediate success" in slowing the coronavirus spread. She emphasized that restrictions of public life remained key to preventing accelerated spread. Schools would reopen on May 4. She encouraged people to wear masks at grocery stores, on public transports, and while shopping.

Retail businesses reopened on April 20. Chancellor Angela Merkel urged people to continue social distancing, wearing masks, and staying at home if possible.

Several authorized and unauthorized protests erupted across Germany on May 1. Gatherings were restricted to fewer than 20 people; however, over 1000 protestors gathered.

A meat-processing plant in Coesfeld was shut down on May 7 after 151 cases of coronavirus were identified. After several meat-packing facilities experienced a higher incidence of COVID-19 infections, new regulations were introduced, including a ban on subcontracting and the introduction of better supervision in employer-provided living quarters.

On May 14, RKI recorded all positive tests, along with probable places of infection. German automakers donated several hundred thousand masks to local hospitals.

Sporadic cases of dispute between Germany and other EU countries over the procurement of masks created a price war. There were political fights within the senate about mask shortages. They cited the US outbidding most of the EU countries to get those masks and PPEs as a cause of the shortage.

As of August 15, the number of daily new cases in Germany increased again after having dropped dramatically in June and July. Germany reported 5-10 deaths per day in August, consistent with the second spike seen elsewhere in the world in places such as Japan, Australia, South Korea, Hong Kong, and several southern US like Texas, Florida, and California.

Case numbers continued to rise throughout September, raising additional concerns due to the approaching 'flu season. It was not clear how the 'flu would influence the second wave of COVID-19. Many EU countries. began implementing a second localized lockdown.

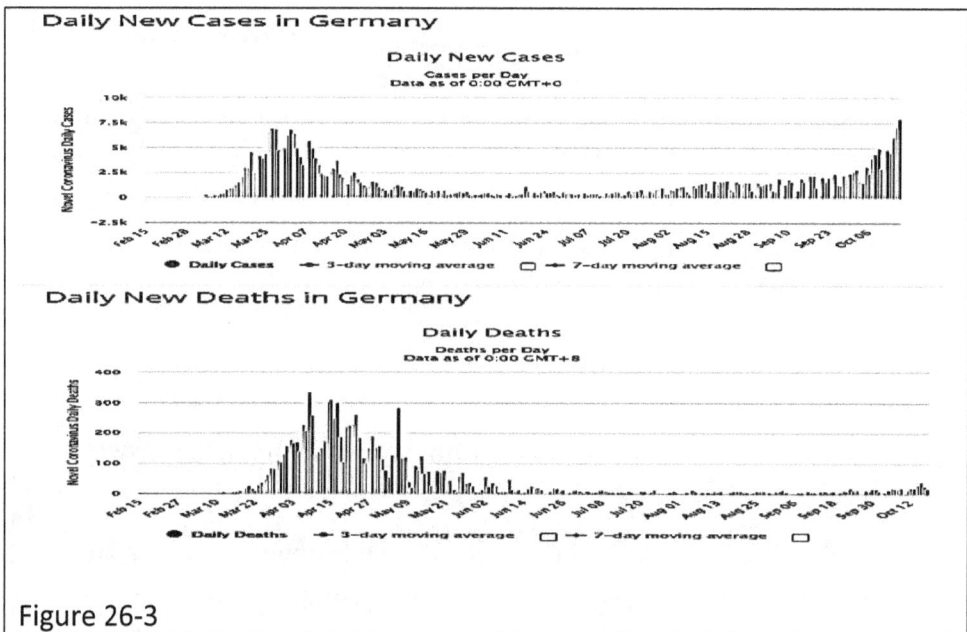

Figure 26-3

GERMANY VERSUS THE UNITED STATES STATS

At the time of writing this article (Sept. 18, 2020), Germany's stats were:

Descriptions	Germany	The US	
Total cases	271,840	6,946,888	
Deaths	9466	203,483	
Total cases/1 M pop	3424	20,961	
Deaths/ 1 M pop	113	614	5 times
Total tests done	14,557,136	96,537816	
Test/1M pop	173,624	291,280	More tests/1 M pop
Population	83.8 M	330.1 M	
NIK NIKAM 09182020	Ref: https://www.worldometers.info/coronavirus		

CONCLUSIONS

Germany faced the same challenges as many other countries did.

They had one of the best national healthcare systems that tested, tracked the contact cases, and quarantined them as needed.

They performed more tests in the initial stages to detect active cases early on than in other countries such as the US, although the US testing numbers were eventually higher.

LESSONS LEARNED FROM THE GERMAN APPROACH.

The low COVID-19 death rate in Germany was a mystery even to the experts from the RKI, like Marieke Degen. It may have been related to how deaths are reported in Germany, with cases either being missed in Germany or the contribution of COVID-19 to deaths being over-estimated in other countries.

Early testing, transparent communication, and professional contact tracing early in the pandemic were clearly helpful in mitigating COVID-19 in Germany. The country had a robust healthcare IT structure and digital health system, and, like S. Korea, Israel, and Singapore, they used phone tracking. However, this approach raised privacy issues in other countries.

Germany tested people with milder symptoms, thus identifying cases might be potentially spreading the disease more quickly than in other scenarios. They had COVID-19 tests ready in January. In comparison, in the US, the FDA relied on large companies to produce their test for COVID-19 detection and did not have tests available until March and April.

Germany had excellent ICU units, which perhaps enabled their sick patients to recover more often than those in other countries. Germany's Digital Care Act allowed for monitoring infected people using telemedicine, limiting infection spread in clinics.

Other experts felt it might be too early to know the real death rate as it is not possible to determine the full extent of the pandemic so early, and numbers may still surge.

Each country should develop or have testing kits before or very soon after a novel virus invades their country. Identifying active cases at the beginning could save thousands overall.

REF:

https://www.worldometers.info/coronavirus/country/germany/

https://medicalfuturist.com/how-germany-leveraged-digital-health-to-combat-COVID-19/

27 NEW ZEALAND'S VICTORY STORY

New Zealand (NZ) is an island country in the southwestern Pacific Ocean. It consists of a large North and the South Island and over 600 smaller islands covering a total area of 268,021 square kilometers. It had a population of 4.886 million in 2018 and was led by Prime Minister Jacinda Ardern during the COVID-19 pandemic. PM Ardern has received accolades from around the world

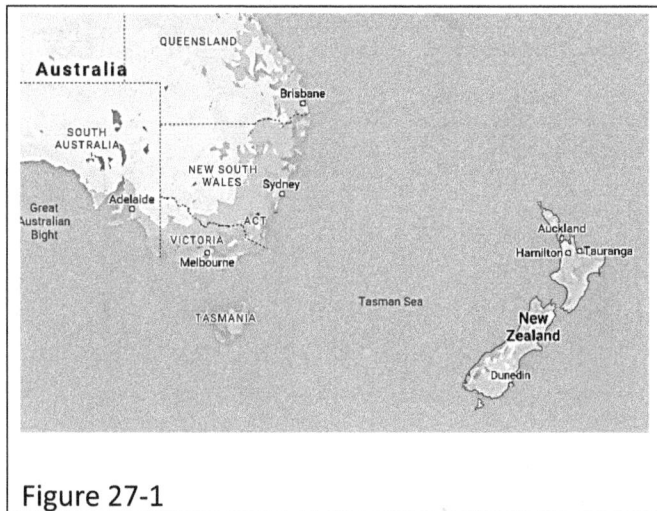

Figure 27-1

for how she contained the coronavirus and almost eradicated it from the islands in June and July of 2020. (Figure 27-1)

As of Sept. 19, NZ had recorded a total of 1811 coronavirus cases with 25 deaths. A remarkable accomplishment. (Figure 27-2)

WORLD / COUNTRIES / NEW ZEALAND
Last updated: September 19, 2020, 18:25 GMT

New Zealand

Coronavirus Cases:
1,811

Deaths:
25

Recovered:
1,719

Figure 27-2

On March 13, PM Ardern instituted a strict self-quarantine for everyone arriving in New Zealand and introduced a complete lockdown a week later. She said they were going hard, and they were going early. They had 122 cases at the point of lockdown. So did Italy and other countries, days to weeks before lockdowns were put into place. NZ's actions demonstrated admirable, apolitical forward-thinking. (Figure 27-3)

In the two weeks after lockdown, the number of cases and deaths steadily declined in NZ with an R-naught value below 1, allowing for the lockdown to be lifted on April 28. People from around the globe praised PM Ardern's model response for its basis in empathy, clarity, and trust in the

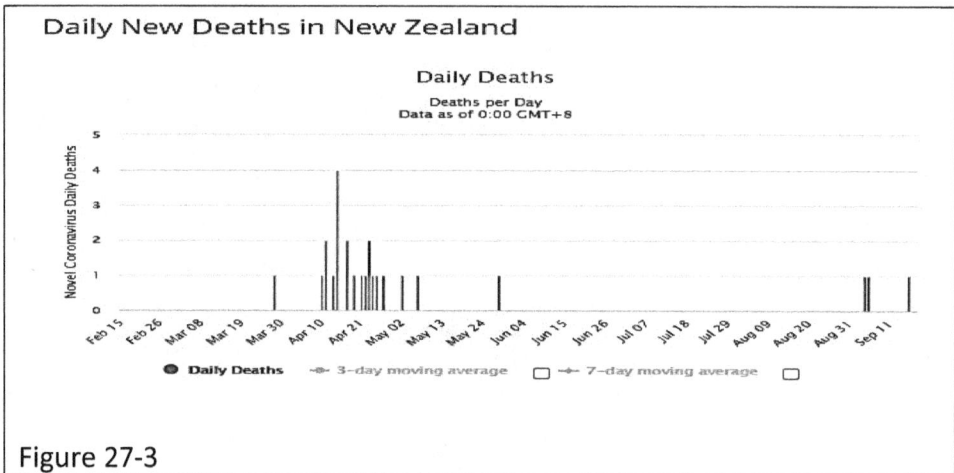

Figure 27-3

available scientific knowledge.

New Zealand is a small island with well-controlled borders and a population of fewer than 5 million people; its population density was about

18.3/square km at the time of the COVID-19 pandemic, while that of New York City's was 10,194/square km.

The government urged people to unite against the enemy, believing that science and leadership had to work hand in hand.

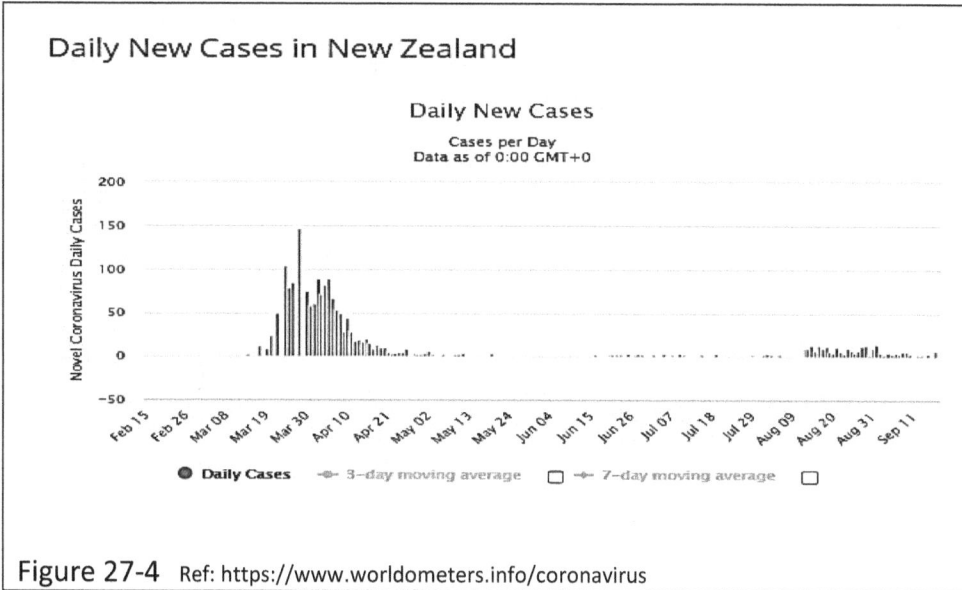

Figure 27-4 Ref: https://www.worldometers.info/coronavirus

I want to emphasize that here we base scientific models on old data and old pandemic models, which might not consider the virulence of a new virus, how it spreads, or the nature of real-time interventions. Hence, scientific projections for pandemics are like weather forecasts.

NZ did a better job of educating the public and explaining the projections with and without lockdown than other countries and propagated the encouraging message to "Be strong. Be Kind" as measures went into place. Arden and her entire cabinet took a 20% pay cut to recognize the impact on the common citizens.

NEW ZEALAND–A SUCCESS MODEL

NZ began aggressively testing the population early in the game when they only had a handful of cases. This paid off overall and is an important lesson for all countries in the event of a new pandemic. (Figure 27-4)

COVID-19-infected patients were quickly identified and isolated for 2-3 weeks, preventing significant community spread. Closing NZ borders and quarantining those who needed to return to the country for 14 days mitigated spread from overseas.

Between April 28 and Aug 15, NZ occasionally reported a case of COVID-19. In the middle of June, two New Zealand sisters who had returned from the UK were in quarantine. Somehow, they left quarantine to visit an ailing family member and later tested positive for COVID-19. PM Ardern put the military in charge of the quarantine program, and they efficiently traced 320 people with whom the two sisters might have had contact and followed up with every one of them to make sure they quarantined and received COVID-19 testing if necessary.

However, as the coronavirus was making a resurgence in the US, the EU, and the rest of the world, even NZ saw an uptick in the number of new cases. Between Aug 15 and September 18, 44 new cases were recorded with a few new deaths.

Most graphs contained in this chapter were sourced from worldometers.info

https://www.worldometers.info/coronavirus/

https://www.axios.com/new-zealand-military-quarantine-border-f4ef1b4e-7019-4433-91fc-9c063ed2b0ae.html

28 SWEDEN AT A CROSSROADS

Sweden is a Scandinavian nation, with Stockholm as its capital. It had a population of 10.23 million people in 2019. Sweden is surrounded by Norway at its west, Finland east, and Poland and Germany to the south.

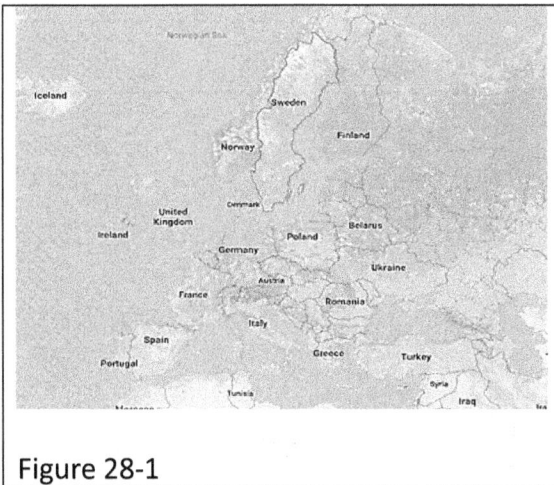

Figure 28-1

Sweden was one of the few countries to adopt a relaxed quarantine measure compared with the strict lockdown seen in other EU countries. It was not clear if Sweden's approach, which was to propagate herd immunity with minimal mitigation measures, was a stroke of genius or a misguided policy that would later haunt them. World experts could blame the lax approach for Sweden's high death rates in comparison with other EU countries with similar populations. (Figure 28-1)

The safety measures used in Sweden were significantly short of a full lockdown. People were advised to follow self-mitigation rules and personal hygiene, and businesses mostly remained open.

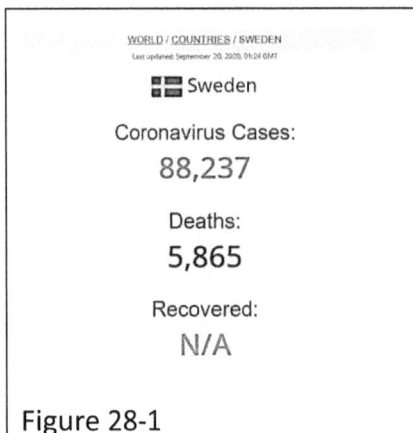

WORLD / COUNTRIES / SWEDEN
Last updated: September 20, 2020, 01:24 GMT

Sweden

Coronavirus Cases:
88,237

Deaths:
5,865

Recovered:
N/A

Figure 28-1

As of September 20, Sweden had 88,237 coronavirus cases and 5,865 deaths. This is just part of the story. If you compare Sweden's numbers to other EU countries, they appear similar; however, the lack of lockdown in the middle of a pandemic baffled world healthcare leaders and did not preserve the country's economy as predicted. (Figure 28-1)

The Bell curves for most countries show a standard response with a brisk rise and gradual fall in the number of cases. However, Sweden's curve remained flat with a periodic surge in new cases with a spike of new cases in June. Even as the number of new cases came down, they never reached the bottom of the curve. I have covered the economic impact elsewhere in the book. (Figure 28-2)

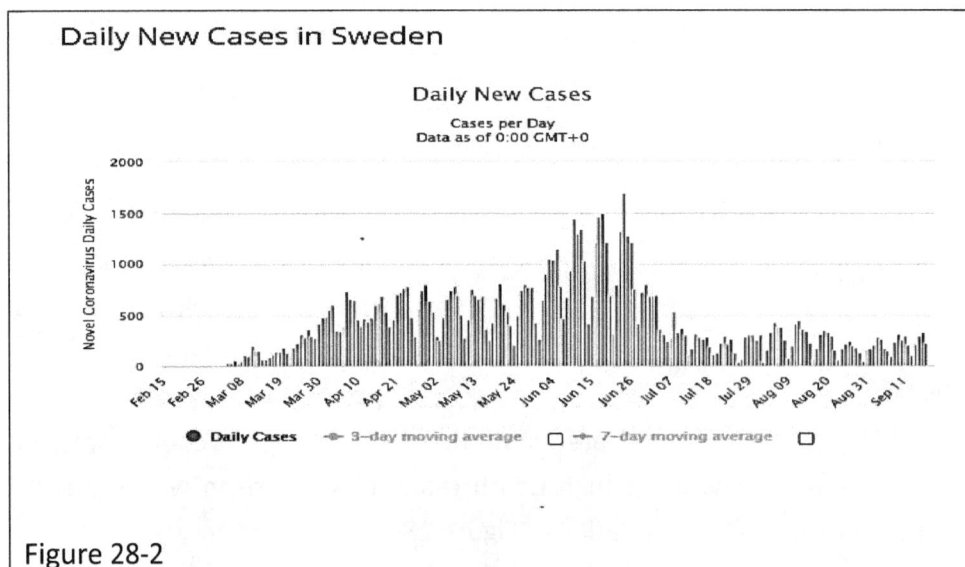

Figure 28-2

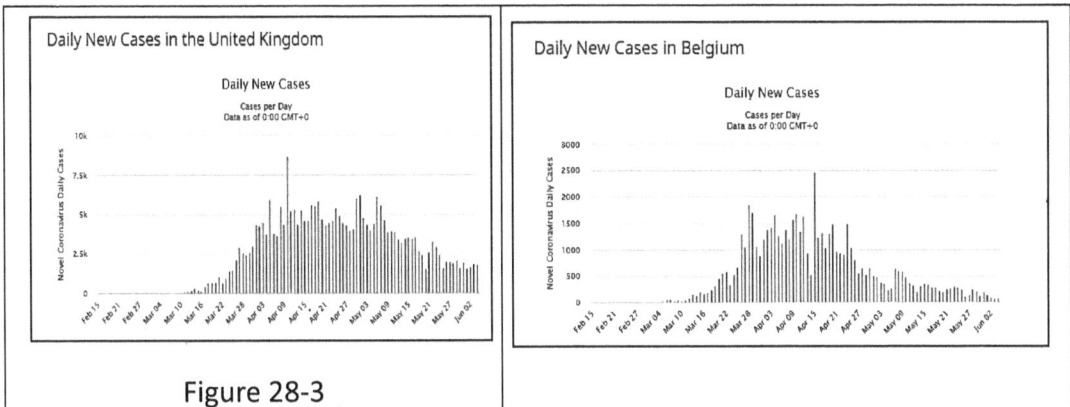

Figure 28-3

Ref: https://www.worldometers.info/coronavirus

Before we dissect the results of Sweden's unparalleled "herd immunity" response, let us first look at the safeguarding measures they had in place:

• Gatherings of over 50 people were banned, and all sports and events were canceled

• Encouraged people to wear masks in public places

• Recommended maintaining a safe distance

• Advised handwashing with soap and water

• Protected the elderly and vulnerable while letting younger people carry on with their lives with precautions

• Kept schools open as they believed children under 14 were at minimal risk and did not spread the virus

• Restaurants and bars remained open

Swedish officials were concerned about the economic impact of a lockdown. The media compared neighboring countries with similar populations that did institute a lockdown to see if Sweden's approach was detrimental.

SWEDEN DEATH RATES COMPARED TO OTHER EU COUNTRIES BASED ON POPULATIONS

I reviewed Portugal, the Netherlands, Switzerland, and Belgium, along with Sweden. These countries had a population of 8 to ten million people, except for the Netherlands, which had a population of 17 million.

First, the number of COVID-19 tests varied significantly between countries. Sweden performed the lowest number of tests, meaning several active coronavirus cases are likely to have been missed.

SWEDEN DEATH RATES COMPARED WITH EU COUNTRIES BASES ON POP.						
	TOT CASES	DEATHS	CASES/1 M	DEATHS/1 M	TESTS	POPULATION
NETHERLANDS	91,934	6275	5363	366	2,042,887	17,132,145
PORTUGAL	68,025	1899	6676	186	2,345,680	10,198,769
SWEDEN	88,237	5865	8725	580	1,393,161	10,094,604
SWITZERLAND	49,283	2045	4509	236	1,252,614	8,649,901
BELGIUM	99,649	9937	8590	857	2,814,376	11,585,939
UK	390,358	19,339	5744	614	21,368,297	67,964,812
Source: worldometer.com (SEP 20, 2020)						
NIK NIKAM, MD. Ref: https://www.worldometers.info/coronavirus						

Figure 28-4

We might conclude that Sweden's unorthodox approach contributed to increased deaths. However, Sweden's death rate was 580 per million population, and Belgium, which instituted a stricter lockdown, had a death rate of 857 per million population. Other EU countries such as Britain (588/1 M), Italy (557/1 M), Spain (580/1 M) also had higher mortality per million. (Figure 28-4)

By early June, it became clear that Sweden was not achieving herd immunity. Despite unrestricted exposure, immunity remained far from the 60-70% rate required for herd immunity.

Sweden's death count data is potentially unreliable depending on how deaths were recorded, e.g., if people who died at home were counted, and if people who died of COVID-19 never received a test. Still, based on the data, Sweden had 5,865 deaths representing and rate of 580 per million. That was much higher compared with the Netherlands, Portugal, and Switzerland. (Figure 28-5)

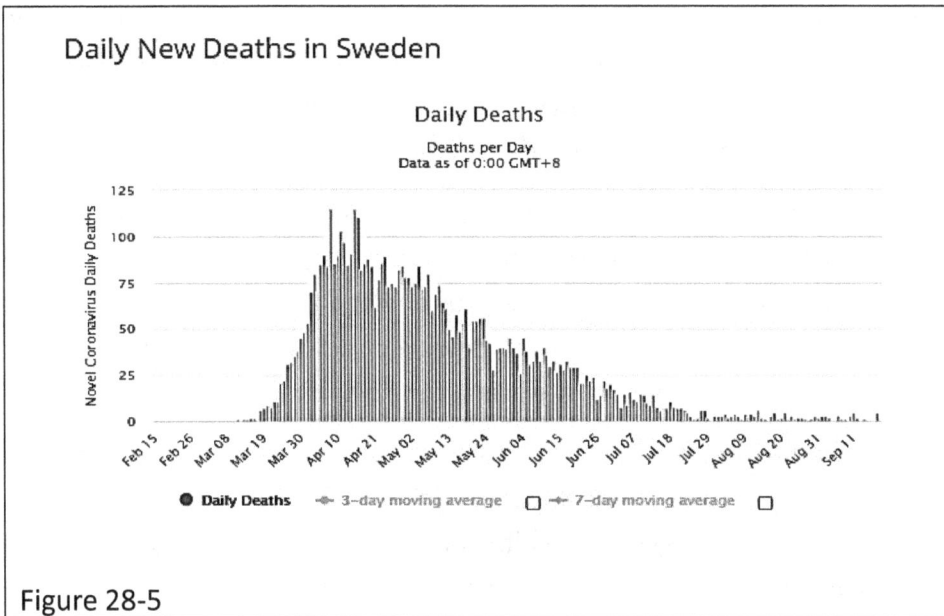

Figure 28-5

Ref: https://www.worldometers.info/coronavirus

DEATH RATES IN VARIOUS EU COUNTRIES

The European equivalent of the CDC (Centers for Disease Control in the US) advises caution when comparing the data from one country with that of another. Belgium reported all confirmed and suspected deaths, whether in the hospital or at home. That might explain why their death rate was so high. Sweden reported only confirmed cases. Finland and Norway reported only hospital-confirmed deaths. Nursing home deaths were not reported in many countries. This makes it difficult to accurately determine the efficacy of any country's pandemic response measures in direct comparison with another country.

Other factors that might have affected the death rates include not admitting nursing home residents with COVID-19 to the hospitals and not transferring anyone over 60 to the ICUs.

COUNTRIES WITH HIGHEST DEATH RATES

Look at the "Deaths" chart from September 20. Italy, the US, the UK, Brazil, Spain, etc. showed much higher mortality per million than Sweden.

However, the flaw in looking at the death rates while not knowing the percentage of positive cases or the number of cases in people aged over 60, the number of people who followed mitigation guidelines, etc. is that it is not possible to accurately determine these death rates without that information.

The Swedish epidemiologist Tegnsell denied that Sweden was deliberately trying to reach herd immunity as quickly as possible. This would have caused a rampant spread of the virus and should have resulted in a vast number of people testing positive; however, if insufficient people were tested, the country's infection rate is inaccurate and meaningless.

The epidemiologist Tegnsell said that looking backward, they should have taken a more middle-ground approach instead of the relaxed mitigation methods introduced. Sweden's health spokesperson, however, said that the total lockdown in other countries was an overreaction and even questioned the wisdom of wearing masks, thinking that masks do not protect people from getting the coronavirus and provide a false sense of security.

In June, a sudden uptick in new cases became a concern for Sweden's healthcare authorities. They felt they should have done something differently, but they were not sure what they could shut down to prevent the slow the coronavirus spread.

Sweden received the recommendations given to other countries that were then preparing to reopen economies after the lockdown. What was the difference? The difference was in the initial 2-3-week period when a lockdown is most needed to slow the viral spread.

Keeping the economy going while the coronavirus swept through the country was one of Sweden's priorities. However, that too fell short of their expectations. Sweden's economy shrank 7% in the year 2020—even more so than during World War II.

When Denmark and Norway opened their borders to their neighbors, they excluded Sweden.

Sweden's political left wanted to appoint a commission to investigate the administration's handling of the coronavirus pandemic, and people's confidence in public health officials rapidly dwindled.

HOSPITAL ADMISSIONS AND THE NUMBER OF PATIENTS IN THE HOSPITALS

Sweden might have had a higher mortality rate that was not accurately reported. However, the number of sick people in the hospitals showed a decline similar to other EU countries that followed strict quarantine measures.

You should view most numbers with some reservations. The data released are likely to be inaccurate.

Additionally, Sweden treated certain older people differently compared with how they would have been treated in other countries, and the reported death rates might have been different. They declined to admit the elderly to the hospitals. Many of them were denied the ICU beds as they wanted to use them for younger people. More than 50% of their deaths came from the care home.

Herd immunity required 60%-70% of the population to have antibodies, and in April, only 17% of the population of Stockholm had antibodies, which was the same percentage observed in London, according to a BBC report. In contrast, 21% of people in New York City had antibodies.

The Swedish administration was at a loss to explain why their approach was not working. Their general trend indicated how severe the effect was, how it progressed over time, and how it compared to those of other countries like Germany or France, where the data were more reliable.

Great lessons have yet to be learned from Sweden's large human experiment, where protocols were based on old concepts and models while largely disregarding scientific recommendations for COVID-19, including recommendations for testing and recording deaths.

One factor that stands out is Sweden's minimal testing. On Sept 20, Sweden had performed 1,393,161 tests. Portugal, with a similar population, had performed 2,345,680 tests. Other EU countries tested 3-4 times more people.

Yes, it was puzzling to see Spain, UK, Italy, Belgium, and France had a much higher number of deaths per million population, but data reporting cannot always be considered reliable. Trust the trends, as they speak to the past, the present, and the likely future.

SWEDEN JULY UPDATE

As of July, Sweden had failed to develop herd immunity or save the economy. Their mortality rate reached 650 per million, and their economy was worse than when they had started. COVID-19 deaths stood at 5,420 for a country of ten million population. The unemployment rate went from 7.1% in March to 9% in May.

Swedish manufacturing stopped when their supply channels outside the country dwindled, underscoring that one country might not be able to exist in a bubble. Most countries are interdependent for materials, human resources, technology, and services.

Sweden's laissez-faire approach has baffled their economic leaders and had not minimized the economic damage

Ref:

PS: Most of the graphs were from the worldometers website

https://www.worldometers.info/coronavirus/

https://www.bbc.com/news/health-53741851

https://www.nytimes.com/2020/07/07/business/sweden-economy-coronavirus.html?

https://cheps.sdsu.edu/docs/Contagion Externality Sturgis Motorcycle Rally 9-5-20 Dave et al.pdf

https://www.france24.com/en/20200916-they-sacrificed-the-elderly-how-covid-19-spread-in-sweden-s-care-homes

29 INDIA: A SLOW BUT STEEP CLIMB TO MOUNT EVEREST!

India, with a population of 1.38 billion, is the second-largest country in the world. In line with many Asian countries, the magnitude of the pandemic in India began mildly in comparison with EU countries and the. (Figure 29-1)

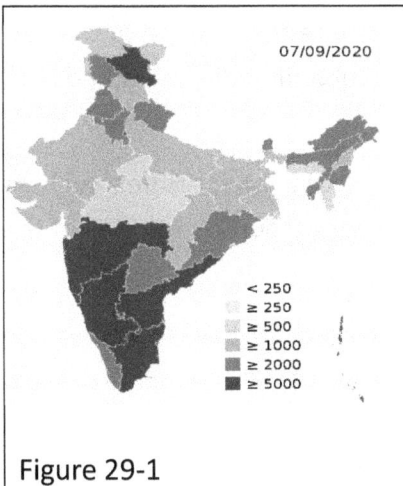

Figure 29-1

India's first case was recorded in Thrissur, Kerala, on Jan. 30. That case originated from Wuhan, China. By February 3, the number of cases rose to 3, all of which originated from Wuhan.

On January 21, India began screening passengers arriving from China and expanded that to include Thailand, Singapore, Hong Kong, Japan, South Korea, Nepal, Vietnam, Indonesia, and Malaysia in February. On March 13, India closed borders from all neighboring countries.

India consistently experienced a slow rise in the number of new cases and had yet to reach a plateau or decline at the time of writing. Wave patterns from other countries indicate that India's decline may be slow and take several months to near the bottom of the curve.

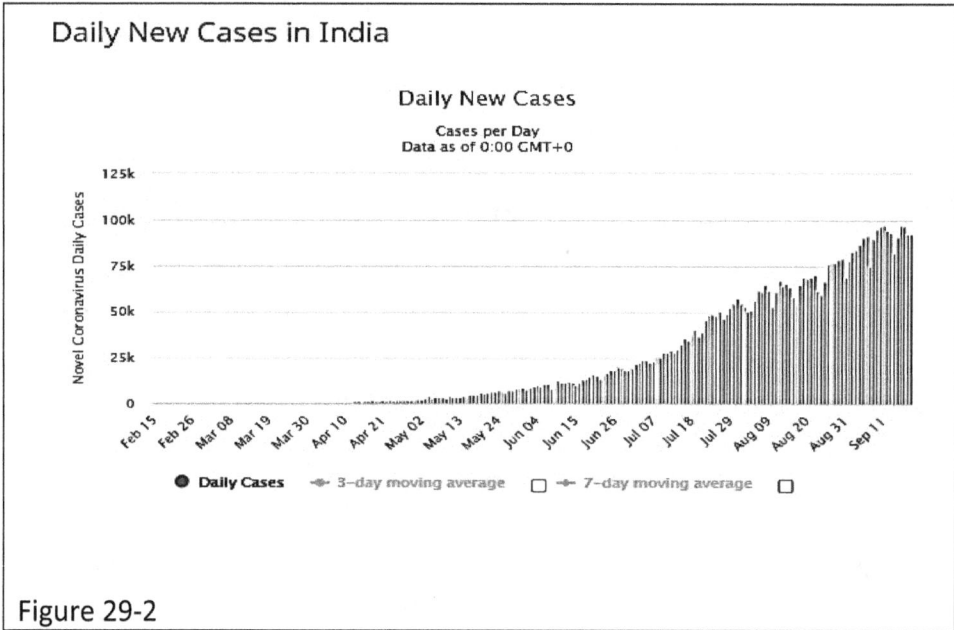

Figure 29-2

In contrast to many other Asian countries where the upslope of the Farr curve lasted for weeks, the upslope of the Farr curve in India has lasted over eight months so far.

India had the largest number of cases in Asia. Major cities such as Mumbai, Delhi, Ahmedabad, Chennai, and Kolkata accounted for 50% of the cases.

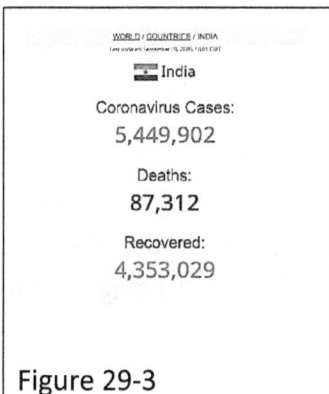

Figure 29-3

India introduced a 21-day lockdown on March 24. On April 14, PM Narendra Modi extended the lockdown until May 3. India announced a second extension of the lockdown for two more weeks on May 1. On May 17, the lockdown was further extended until May 31st.

Many businesses and educational institutions closed, and tourist visas were suspended.

Looking at India, the second-largest populated country, some experts were concerned about the impact of the virus on the population and the world at large. Proponents of the sustained lockdown argued that the doubling rate of new cases went from six days (April 6) to eight days (April 18) and that lockdown was necessary to save lives. Others focused on the colossal economic loss and devastation that would result on top of a GDP had already been steadily declining.

On 12 March, after the World Health Organization (WHO) declared the coronavirus a pandemic, the Indian stock market suffered its worst crash since June 2017. The BSE SENSEX dropped 2,919 (8.18%) points, reaching its lowest in 23 months.

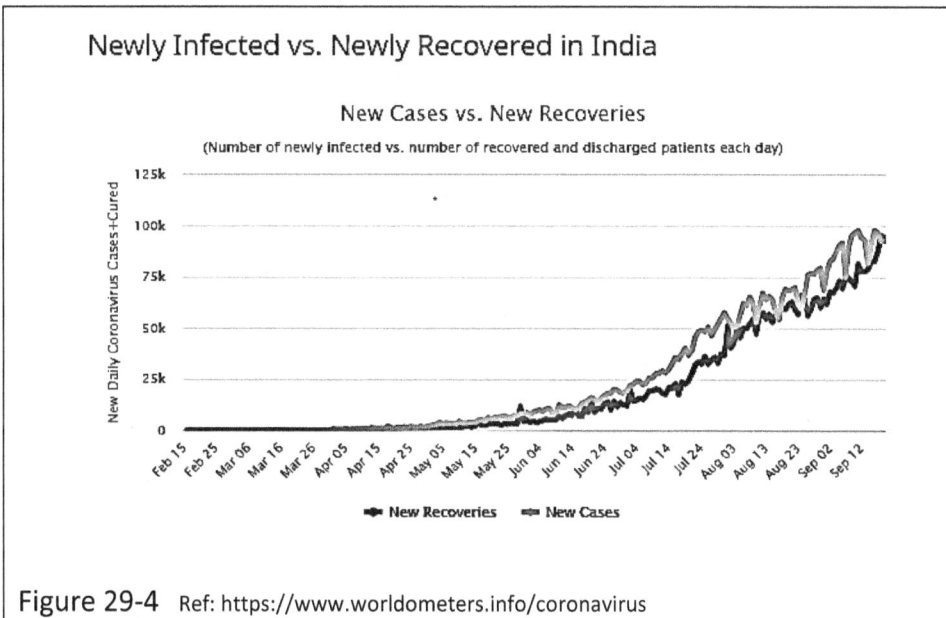

Newly Infected vs. Newly Recovered in India

New Cases vs. New Recoveries

(Number of newly infected vs. number of recovered and discharged patients each day)

Figure 29-4   Ref: https://www.worldometers.info/coronavirus

On March 12, PM Narendra Modi introduced an economic package comprising Rs 20 lakh crore, nearly 10% of India's GDP. In comparison, the Unites States allocated 10% of its GDP and Japan set aside 21% of its GDP toward corona relief.

(For those wondering: Rs 20 lakh crore is $300 Billion, which is 10% of India's GDP that is slightly shy of USD 3 trillion. 1 crore = 10 million. 1000 million = 1 billion. 1000 billion = 1 trillion.)

The package was geared toward the workforce, farmers, taxpayers, MSMEs, and cottage industries.

On March 13, Modi proposed that SAARC nations (Nepal, Maldives, Sri Lanka, Bhutan, Bangladesh, and Afghanistan) jointly fight the coronavirus pandemic and pledged a $10 million emergency fund for those countries.

On March 17, the Indian government introduced social distancing until March 31. On March 22, a "COVID-19 Economic Response Task Force" was established. Face masks were mandated around the same time.

The Government divided the country into three zones:

• Red zone (Hotspots) – districts with a high doubling rate and a high number of active cases

• Orange zone (Non-hotspots) – districts with fewer cases

• Green zone – districts without confirmed cases or no new cases in the past 21 days

The number of cases started climbing in March and mostly originated outside of the country.

On March 24, Modi announced ₹150 billion in aid for the healthcare sector. Money was allocated for PPE, ICUs, ventilators, developing testing, and training medical workers.

On 28 March, the Prime Minister's Citizen Assistance and Relief in Emergency Situations Fund (PM CARES Fund) was set up to provide relief to the population. Many large businesses contributed to the relief fund. On May 13, they allocated a sum of ₹3,100 per person.

On March 31, a religious program in Nizamuddin West Delhi, the Tablighi Jamaat event, attracted thousands of people despite a ban on gatherings over 200 people. The event resulted in 4291 positive cases representing 29% of India's confirmed cases. Twenty-two thousand people had to self-quarantine after close contact with those who attended the event.

On April 4, Modi banned the export of hydroxychloroquine to ensure there were adequate supplies for domestic use. However, under pressure from US President Trump, India partially lifted the ban on humanitarian grounds.

In mid-April, the Indian Council of Medical Research (ICMR) invited hospitals and institutions to take part in convalescent plasma transfusion for coronavirus patients. Early on, Delhi reported excellent results in four people who had received convalescent plasma. In June, a clinical study began on its safety and efficacy. In September, India reported mixed results in 450 patients, covered in the Treatment chapter.

On April 14, Modi instructed people to follow seven steps: Use homemade masks, take care of elderly people, protect jobs, help the poor and needy, follow the guidelines set by the Ministry of AYUSH to improve immunity, and download the "Aarogya Setu" app to track your health.

India had close to 40,000 ventilators, of which 8,432 were in a private healthcare setting. Several private companies such as DRDO, ISRO, and auto manufacturers joined hands to produce portable ventilators.

By May, India was producing 200,000 items of PPE and 250,000 N95 masks per day. By the second half of May, India was the world's second-largest producer of PPE.

By May 2, active cases had risen to 10,000, topping 50,000 by May 7, and exceeding 100,000 cases by May 19. In the ensuing eight days, the count climbed to 150,000 cases.

By May 10, India had already registered more than 2000 deaths.

India was among the first countries to evacuate its citizens from China during the Wuhan epidemic. In early May, the "Vande Bharat Mission" evacuated 14,800 Indian citizens from across the globe.

The Ministry of Electronics and Information Technology launched a smartphone app called "Aarogya Setu" for contact tracing to contain the viral spread.

The National Institute of Virology (NIV), Pune, and 15 other private labs performed tests in India initially, and by early March, over 65 labs were capable of testing for the virus. The government paid for two tests for everyone.

With the help of the NIV, India became the fifth country to isolate a pure sample of the coronavirus that could eventually lead to the development of new drugs and vaccines.

In May, the NIV introduced another antibody test kit ELISA for rapid testing, capable of processing 90 samples in a single run.

India faced the same challenges as other countries and had to abandon several tests that turned out to be unreliable.

The list of drugs tried in India, besides hydroxychloroquine, included anti-swine 'flu drugs, anti-HIV drugs, and Ayurvedic medicines.

Several companies, including the Serum Institute of India, Zydus Cadila, and Bharat Biotech, were in the race to develop and test a vaccine for the coronavirus. As early as March, they began testing vaccines in animals.

Asian Development Bank estimated that the coronavirus outbreak might lead to an economic loss to a tune of 29.9 billion US dollars in India.

The plight of the migrant workers, who were the primary labor force in many states, turned sour; they were suddenly out of a job and had no money. They had no straightforward way to return to their villages and were vulnerable to both harsh weather and the coronavirus as they tried to return home on foot, covering hundreds of kilometers. Thus, the supreme court ruled that the central and state governments had an obligation to provide food, shelter, welfare, and transportation to the migrant workers. Individual states pledged millions of dollars as an immediate relief package to support all those who were out of work.

In Wuhan and NYC, the case downslope took twice as long as the upstroke, and as of Sept, if India follows the trend, it could be 6-12 months before case numbers begin to decline. Until then, the best advice is to continue mask use, social distancing, handwashing, and staying safe at home and work as much as possible.

The Indian Bollywood industry, sports, and recreational activities came to a dead halt. Many artists had to find alternative ways of entertaining their fans using live streams and YouTube media. The internet and social media, to a certain extent, kept the music industry alive as artists streamed media from India across the globe. It was safe, involved no travel, and they could deliver from their homes. All was not lost!

THE TOP THREE NATIONS—COVID-19 RACE, THEIR STATS, AND OUR FATES!

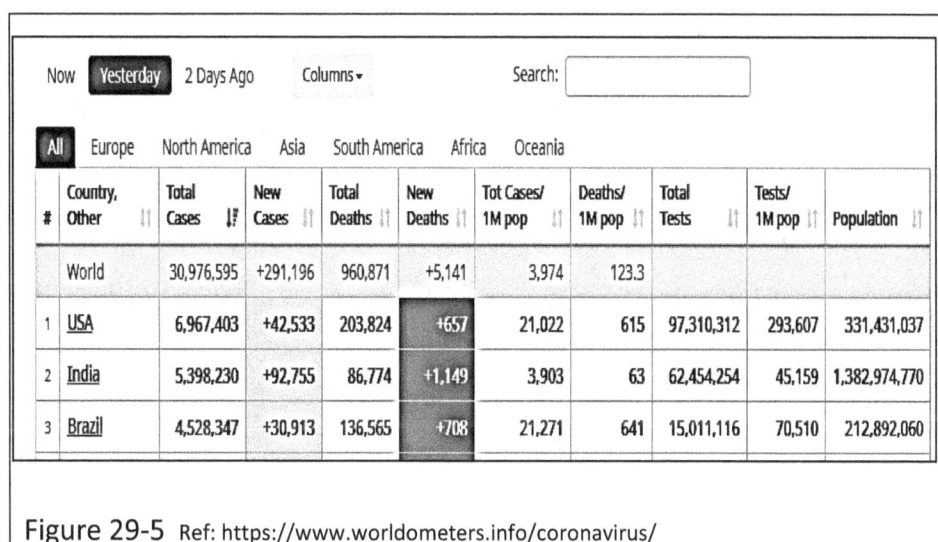

#	Country, Other	Total Cases	New Cases	Total Deaths	New Deaths	Tot Cases/ 1M pop	Deaths/ 1M pop	Total Tests	Tests/ 1M pop	Population
	World	30,976,595	+291,196	960,871	+5,141	3,974	123.3			
1	USA	6,967,403	+42,533	203,824	+657	21,022	615	97,310,312	293,607	331,431,037
2	India	5,398,230	+92,755	86,774	+1,149	3,903	63	62,454,254	45,159	1,382,974,770
3	Brazil	4,528,347	+30,913	136,565	+708	21,271	641	15,011,116	70,510	212,892,060

Figure 29-5 Ref: https://www.worldometers.info/coronavirus/

The US experienced the greatest number of coronavirus cases in the world, reaching almost 7 million at the time of writing. The US also had also performed 97,310,312 tests, representing 27% and the greatest number of tests performed in a single country. The US death rate was also high at 615 death per million, with 203,824 deaths. As of Sept. 20, the US was recording 40,000+ new cases per day.

India, with a population of 1.3 billion people, came in second in terms of many stats. As of September 20, India had 5,394,230 cases, with 92,755 new cases per day—the highest number of cases per day in the world.

If a similar trend were to continue, India might overtake the US as the country with the largest number of coronavirus cases recorded 40 to 60 days after Sept.

However, the total deaths in India compared to the number of infected people tells a different story. India's death rate of 63 per million is 10% of that seen in the US. However, this number is vague and may not accurately reflect reality. When you look at the number of deaths versus active cases (USA deaths at 203,824 per 6,967,403 cases versus India at 86,774 deaths per 5,398,230 cases), the percentage of deaths ends up being 3.1% versus 1.6%. These numbers are vastly different than what is observed with deaths per million.

This highlights the difficulty in using numbers to assess outcomes. The way numbers are used can be chosen in any manner to suit one's narrative. Whichever way you look at it, the mortality of COVID-19 in the US is twice that of India.

Brazil was in third place with 4,528,347 coronavirus cases and 136,565 deaths. When you compare the percentage of death related to the number of active cases, Brazil's mortality is similar to the US, with a 3.01% and 3.1% mortality rate, respectively.

In conclusion, the coronavirus is still here, and we are still mostly in the first wave. Now, as we approach the fall, the 'flu season presents an additional complication due to the risk of infection with both viruses with potentially devastating implications. Therefore, 'flu vaccinations and continuation of the self-mitigation rules are strongly recommended until a Knight in shining armor comes to the rescue—namely, a reliable COVID-19 vaccine!

I have hope that we will have a vaccine by 2021 for some much-needed relief.

https://economictimes.indiatimes.com/news/economy/finance/latest-stimulus-package-among-largest-in-the-world/articleshow/75701976.cms

https://www.cnbc.com/2020/03/26/coronavirus-india-needs-a-support-package-larger-than-20-billion-dollars.html

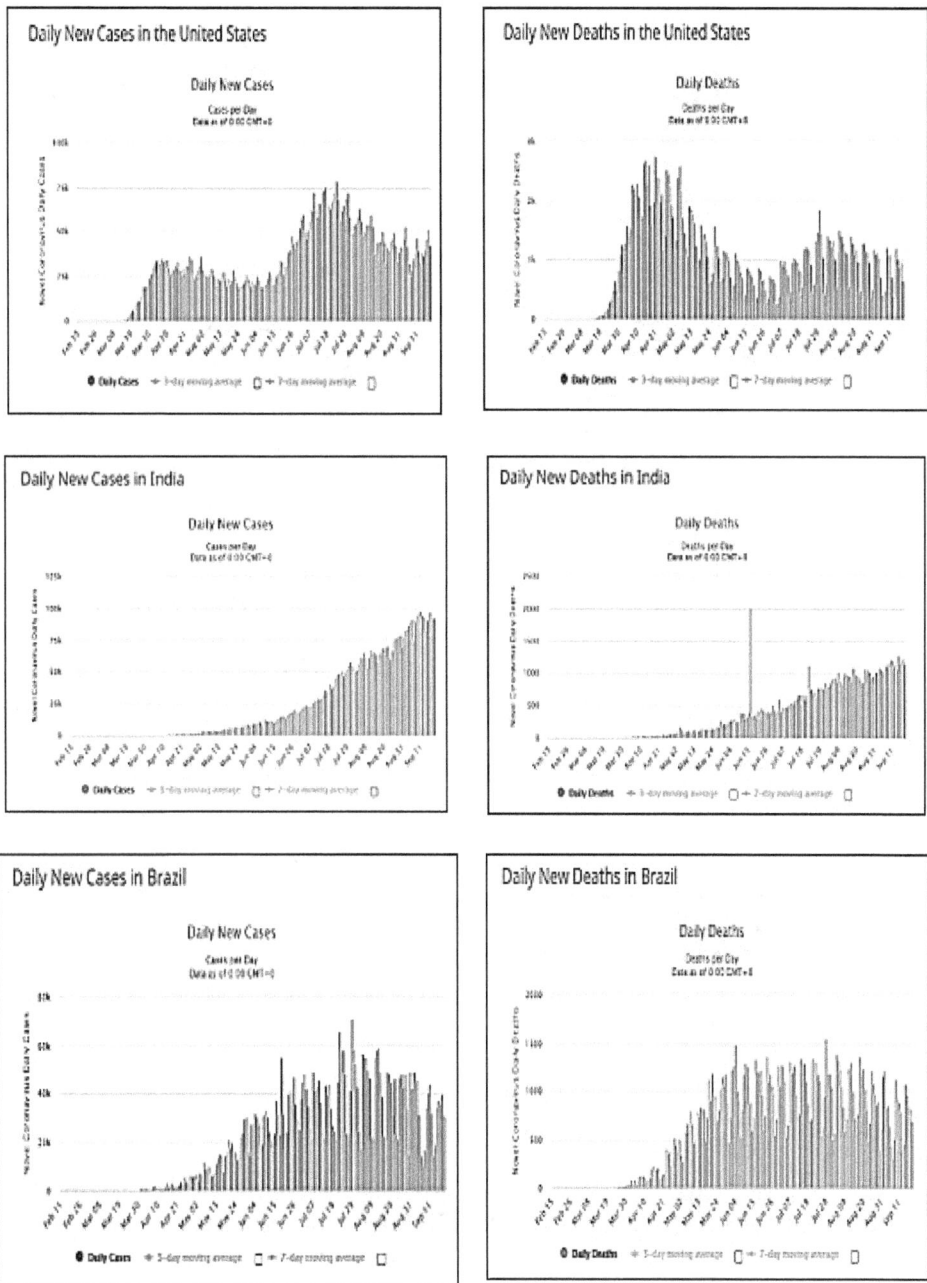

Figure 29-6 Ref: ttps://www.worldometers.info/coronavirus/

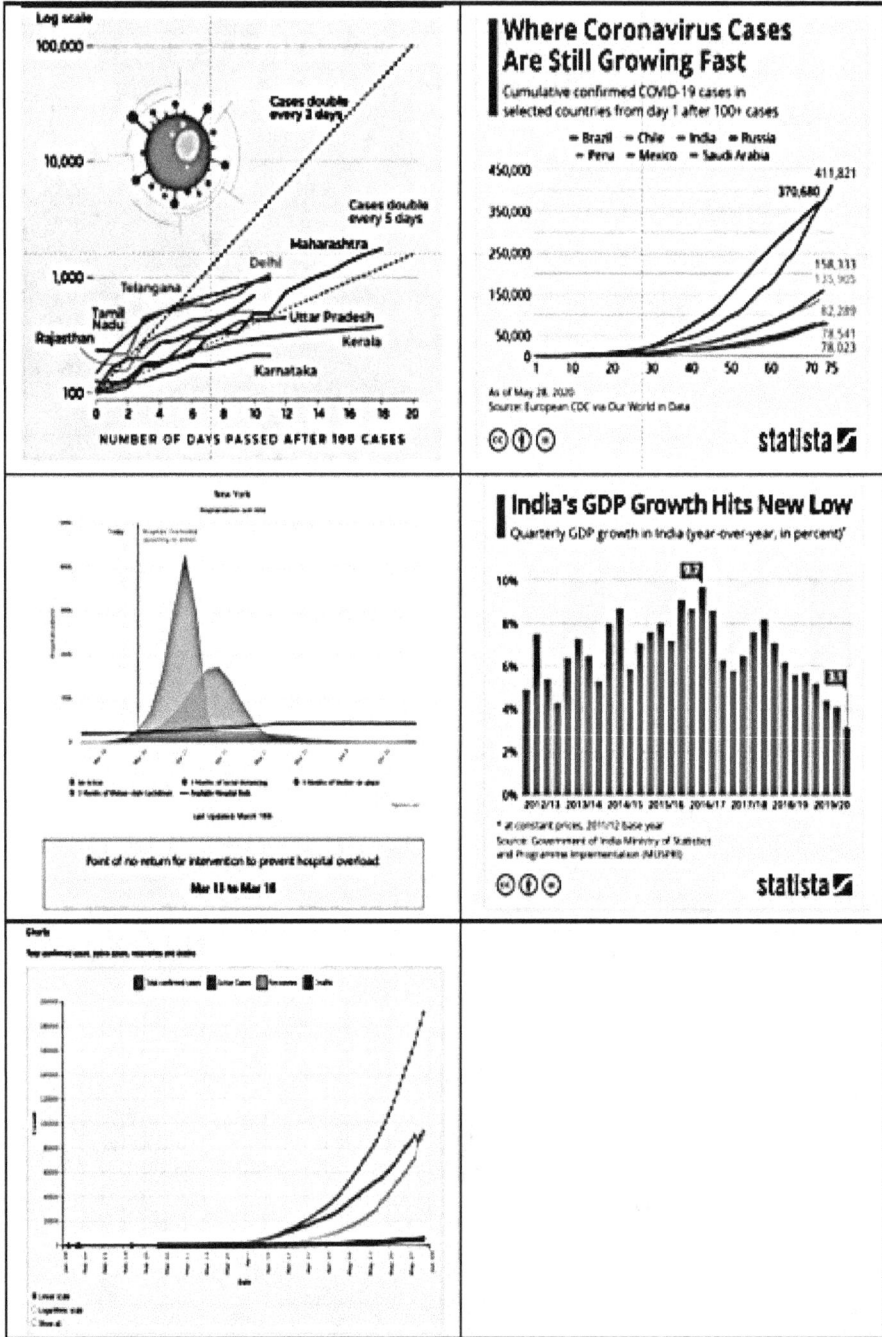

Ref: ttps://www.worldometers.info/coronavirus/

30 TURKEY AND IRAN–TWO TALES FROM THE MIDDLE EAST

Turkey and Iran are rich middle east countries with substantial cultural heritage, oil, and natural resources. They have a population of 84 and 83 million, respectively, and have advanced medical services. Turkey is the world's 17th-largest country by population.

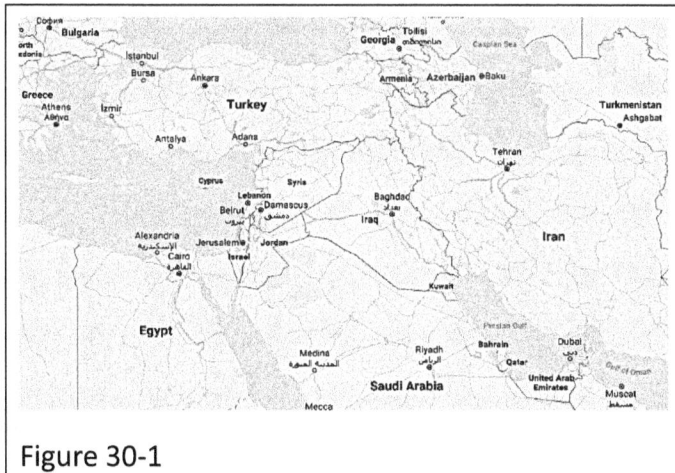

Figure 30-1

The coronavirus pandemic hit Turkey in the middle of Feb. 2020, right after Wuhan had recovered from the pandemic. (Figure 30-1)

COMPARISON

Turkey and Iran have similar population sizes and had similar numbers of active coronavirus cases.

NEW CORONAVIRUS CASES

On Sept 21, Turkey had 302,867 coronavirus cases, with 7,505 deaths. They had performed 9,269,482 diagnostic tests. Turkey's death rate was 289 per million, and its case rate was 3583 per million. (Figure 30-2)

	POP	CASES	DEATHS	TESTS	DEATH/1M	CASES/1M
TURKEY	84,282,595	302,867	7505	9,269,482	289	3583
IRAN	83,786,916	422,140	24,301	3,746,629	89	5012

Figure 30-2

Iran had 422,140 coronavirus cases, with 24,301 deaths. The death rate was 3.5 times that of Turkey. They also had 120,000 more coronavirus cases during the same period. Iran performed 40% fewer tests than Turkey.

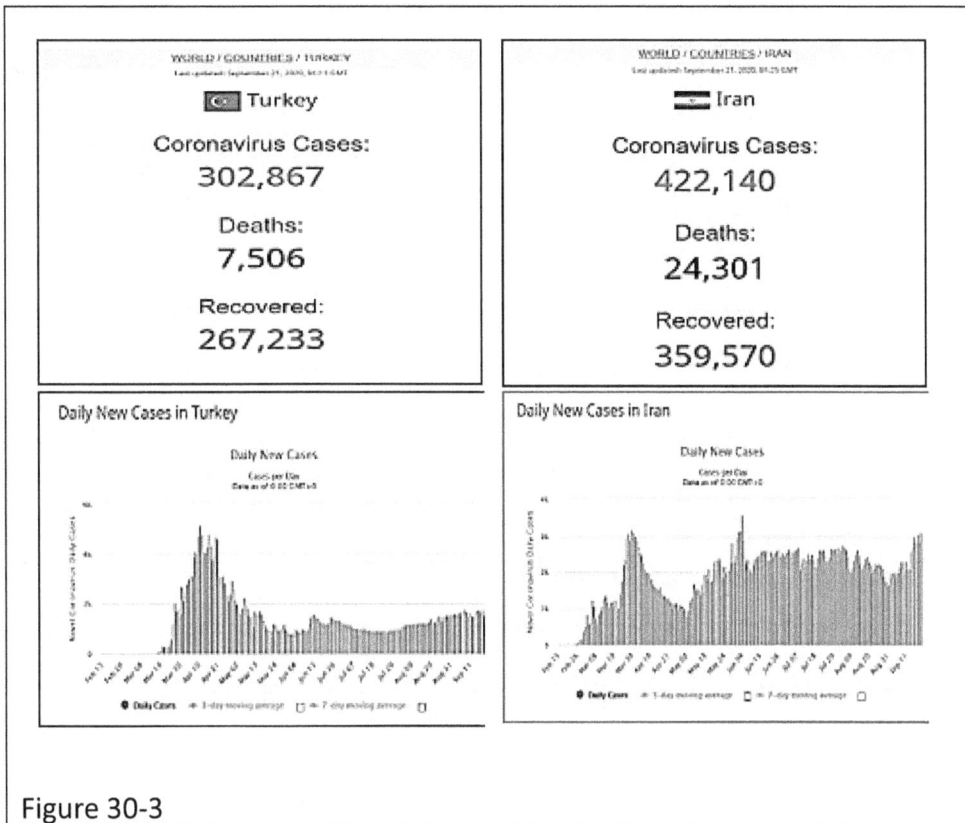

Figure 30-3

Ref: ttps://www.worldometers.info/coronavirus/

DEATH RATES AND GRAPHS

The most striking difference was in the number of people who died from the coronavirus in Iran compared with Turkey: 24,301 and 7505, respectively.

In Turkey, there was a sharp rise in the death count followed by a steady decline, though it did not reach the bottom of the curve. A steady death number was maintained from June to August, which suddenly rose again toward the end of August and September, like the second spike noted in other parts of the world.

Iran also noted a quick rise in the death count beginning toward the end of February 2020. A more gradual decline followed until the middle of May. Between mid-May and August, Iran had a sharp rise in the death count, which was higher than in the first cycle. The death rates came down again yet began to spike again at the beginning of September. No other country had experienced a third wave at this point.

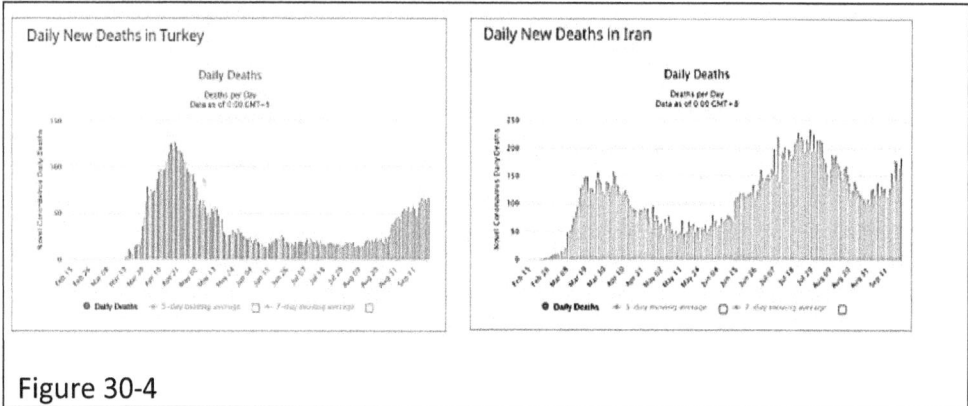

Figure 30-4

Ref: ttps://www.worldometers.info/coronavirus/

HOW WERE THE TWO COUNTRIES' APPROACHES DIFFERENT?

How did Turkey cut the death toll by 45% compared with Iran?

In Turkey, doctors did not advise people with fever, sore throat, or cough to take antipyretics and stay at home. Instead, they invited them to the hospital and immediately started them on chloroquine without waiting for the results from the lab. Besides chloroquine, they added azithromycin to the regimen. Another drug they used widely was favipiravir, which was initially used in China for intubated patients. (Figure 30-4)

Using high-pressure oxygen in patients with respiratory difficulties in the early stages, Turkey was able to reduce the number of deaths related to respiratory failures. They also put their sick patients in the prone position to help drain lung secretions.

Koca, a spokesperson for Turkey, said that these early treatments had been effective in reducing mortality. Since the states controlled most of the official media in both countries, the data must be taken at face value in this context.

Turkey, while fighting its coronavirus challenges along with ongoing border disputes, was able to ship PPE to many other countries.

In February, Turkey cut off travel from Iran as it was the current hot spot in the middle east. Turkey also evacuated nationals from Europe, where the coronavirus epidemic was in full swing.

Turkey became a hot spot in the middle east as it was an international hub for many airlines traveling from the West and Europe to Asia and other countries. Their citizens returning from other hot spots also might have brought the coronavirus along with them.

Turkey did not mandate a complete lockdown but introduced complex curfew rules and age restrictions. It is not clear how effective their curfew rules were in reducing coronavirus spread and mortality.

Turkey invested billions of dollars in its healthcare infrastructure, hired the best scientists, provided free healthcare to all, and believed it had the pandemic under control. Turkey also had a higher ICU bed capacity in its major cities than in other countries.

Turkey reduced the R-naught from 1.56 on May 13 to 0.72 by June 9. Throughout, Turkey kept its ICU mortality below 1% with aggressive early intervention with drugs and oxygen.

The Turkish Medical Association was unconvinced. They felt that they left the borders open too long after the epidemic started. Yet, Turkey had received praise from the WHO for its approach in keeping the death rate low, with 4279 deaths compared with 34,000 in Italy.

IRAN'S CHALLENGES

Iran first detected coronavirus cases as early as Jan 22; however, the regime said there was no sign of coronavirus until four weeks later. The actual Iranian coronavirus crisis began in the third week of February when the regime admitted they had coronavirus cases.

Even though the initial response was slow, the authorities resorted to containment as the epidemic spread throughout the country.

Iran still had flights flying to China, and even Wuhan, in Jan and Feb.

By mid-April, strangled by the economic crisis and the free-falling oil prices, Iran was forced to jump-start its economy too early.

Soon after the country opened for business, its active coronavirus cases rose rapidly, creating a second wave amidst unrest. This created a rift between the general population and regime officials.

Even before the coronavirus crisis, the regime had battled a violent protest in Nov. 2019.

Experts believe that the celebrations of the 41st anniversary of the revolution and parliamentary elections augmented coronavirus spread across the country. On Feb 19, Iran acknowledged that the country was experiencing coronavirus deaths.

Iraj Harirchi, the deputy health minister, tested positive for the coronavirus. By early March, 8% of the Iranian parliament had tested positive for the coronavirus.

Iran experienced a severe PPE shortage yet refused humanitarian help from the US.

Iran's official numbers may be a gross underestimation of the actual number of new cases and deaths.

After the second resurgence, a lockdown was reinstated in the southwestern part of Iran. Several Iranian leaders admitted that the people were unhappy and expressed unrest.

LESSONS FROM IRAN'S MODEL

One lesson for all other countries when planning to re-open the economy is to look out for the beginnings of a second wave and take aggressive steps to identify, quarantine, and contact-trace any new cases.

The second lesson is that authorities need to be transparent as people are unlikely to be compliant if any cover-ups or inconsistencies are apparent. That was a wake-up call for leaders all over the world who were pushing to open their economies despite persistent coronavirus infections in their countries.

COVID-19 PANDEMIC 2019-2020

Failing to take strict measures could result in a devastating second wave and push countries into a state of chaos, with people revolting against conflicting guidance and untrustworthy leadership.

Ref:

https://www.hurriyetdailynews.com/turkish-model-proves-effective-in-COVID-19-treatment-154220

https://www.jpost.com/international/how-did-turkey-get-the-most-coronavirus-cases-in-the-middle-east-625209

https://www.bbc.com/news/world-europe-52831017

https://www.bloomberg.com/news/articles/2020-06-09/delhi-overwhelmed-by-COVID-19-cases-after-city-eases-lockdown

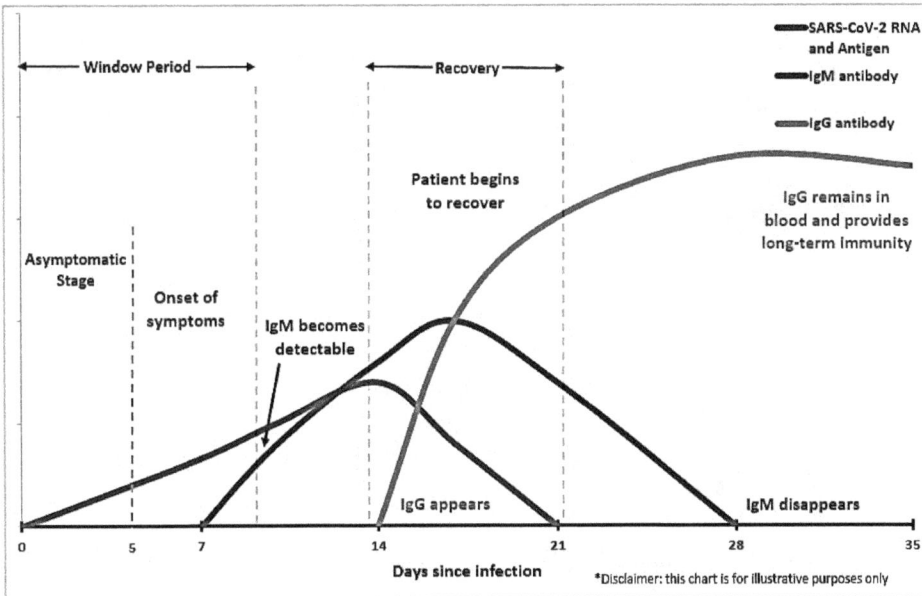

31 ECONOMIC EFFECTS OF COVID-19

THE FALL AND RISE OF THE GLOBAL STOCK MARKET INDICES

As the COVID-19 pandemic spread across the world, penetrating more than 200 countries, forcing people to flee their workplaces and go into lockdown, the global markets were on a colossal downward spiral, the likes of which people had not experienced in decades.

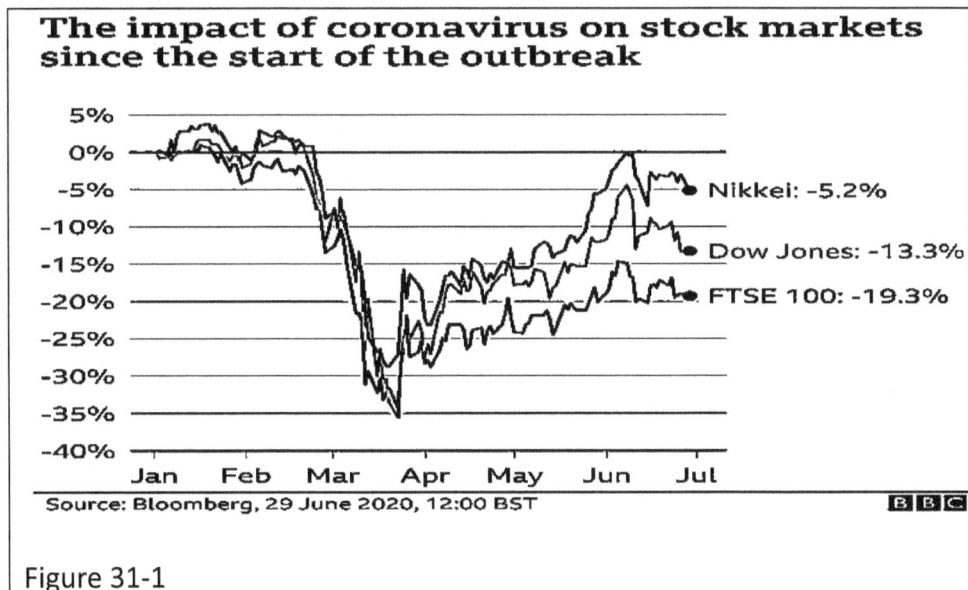

Figure 31-1

The COVID-19 pandemic started at the beginning of Jan 2020, and by the third week of March, the global indexes had dropped 30-35% from their peak in the middle of Feb. The US Dow lost 35% of its value in a little over 30 days. Similar drops were observed in the UK's FTSE (-35%) and Japan's NIKKEI (-30%). Many other countries had similar drops in their stock indexes.

On March 23, the DOW dropped to 18,592 points, representing a 34% drop from the February peak after the coronavirus economic stimulus bill failed for the second time. (Figure 31-1)

After March 23, a brisk recovery started, though it did not progress as steeply as the fall. Despite rapidly spreading coronavirus across the globe and increasing deaths, the US technology index (QQQ) made an unprecedented bounce back, surpassing its February peak on June 2 and continuing an upward trend throughout the progression of the pandemic. In August, while US Congress was stalling a second stimulus package, Pres. Trump signed an executive order extending $400 per week to each unemployed American who had been earning less than $100,000 per year. A payroll tax holiday was extended until the end of December, and a no-eviction clause was extended for renters. The order also put student loan payments on hold until the end of December.

INDUSTRIES GOING FROM BAD TO WORSE

The most affected industries included airlines, the hospitality industry, and retail businesses, to name a few. Because of air travel restrictions, airline stocks took a much bigger hit than the main indexes and lost over 80% of their value from their February peak levels. Many retailers shut their doors permanently. Between March and July, shopping malls across the US turned into ghost towns.

The live entertainment industry went into hibernation. Some artists turned to online performances to a fraction of the audience at a fraction of the cost. Along the same lines, in March, most cultural institutions and theme parks bolted their doors indefinitely.

Retailers such as JC Penny, Nordstrom, Macy's, and Kohl's lost $12.3 billion in market caps. J. Crew, Neiman Marcus, and JC penny filed for bankruptcy. Some stores ramped up their online businesses, which surged 40% over the previous year for some, such as Walmart.

The pandemic pushed restaurant businesses into starvation as their businesses suffered a major setback due to closure orders at the onset of lockdown. Catering services all but vanished, and so did the galas, meetings, functions, graduations, and parties that helped to sustain the food industry. Carry-out became the restaurants' last lifeline, not knowing when they would resume their normal services. Many hourly employees lost their jobs permanently. People were encouraged to support their local restaurants by ordering takeout food.

Tourism became a distant wishful dream of the past. Travelers faced many challenges, including gathering restrictions, lengthy delays, missed connections, fear of getting sick on the plane, cost, and fear of quarantine at the ultimate destinations. Cruise liners lost share reduced by 70%-80%, and it is not clear when they will see the shining seas again.

SILVER LINING AMONG SHADOWY CLOUDS

Despite the ravaging pandemic, some industries made a fortune. Since most people were on lockdown, they had to rely on delivery services for food and many other items. Amazon was the biggest winner in this pandemic as it was ready and had the infrastructure to deliver the services on time. Amazon's stock not only recovered from all the losses but also continued to make new highs frequently. Its stock almost doubled its value from its March low point. Similarly, grocery stores, delivery services, and shipping industries made windfall profits.

Large supply shortages were expected to affect several sectors due to panic buying, increased demand for goods to fight the pandemic, disruption of factories, and transportation of goods, among many other factors. The most notable shortages were that of PPE and pharmaceutical drugs.

NOT EVERYTHING WAS AS PROMISING AS THE STOCK MARKET

Despite a raging stock market with irrational exuberance, the rest of the economic data were neither convincing nor promising. Unemployment numbers were still in double digits in the US, and as late as Aug 5, the government was debating a second economic stimulus package worth $1 trillion (with a capital T!).

FINANCIAL FALLOUTS

In response to the coronavirus pandemic, the central banks of many countries dropped their interest rate to make borrowing cheaper. Sometimes, rates dropped to or below zero. That had a rippling effect on fixed interest rate investments, where the returns plummeted, affecting many people's retirement funds

Many professionals, including doctors, had to take a 10-20% pay cut, adding to the global unemployment crisis. Unemployment in the US hit 14.3% during the epidemic's peak. Many multi-year job expansions melted like a glacier struck by lightning. The government bailed out companies to keep their employees on the payroll. Experts warned it might be many years before the employment number would return to pre-pandemic levels.

Many hourly workers could not pay rent or put food on their tables. Highly skilled workers were lost due to illness. You can always get another job, but you cannot replace a person. A University of Chicago study estimated that moderate social distancing would result in $8 trillion in mortality benefits.

RECESSION RISKS

The economy is measured as the percentage change in Gross Domestic Product (GDP) or the value of goods and services produced, typically over three months or a year.

But the IMF predicted the global economy would shrink by 3% in 2020, perhaps representing the worst decline since the 1930 Great Depression. IMF also forecasted the global growth rate at 5.8% in the post-pandemic year 2021. (Figure 31-2)

Normally, a growing economy means more wealth and more new jobs.

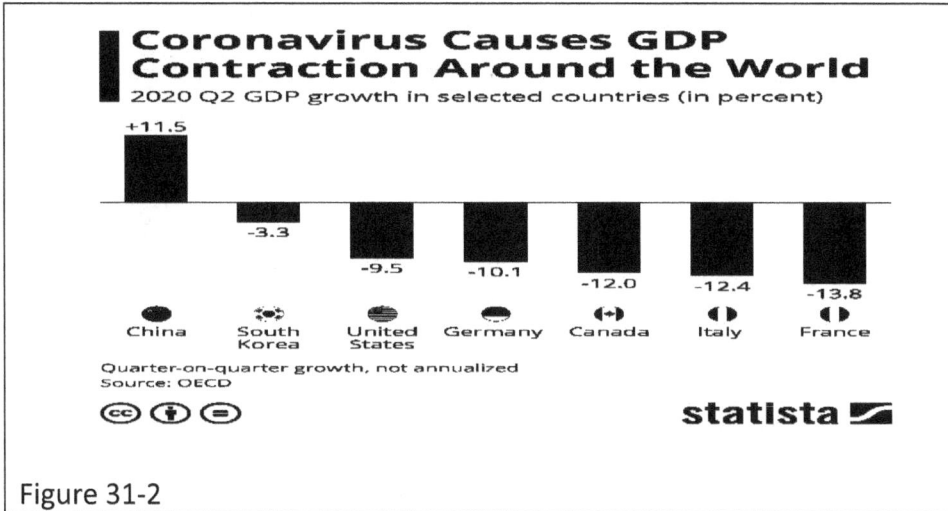

Figure 31-2

As of September 7, with the COVID-19 pandemic showing no signs of slowing down and the vaccine still in the works, the economic toll was compounding month after month.

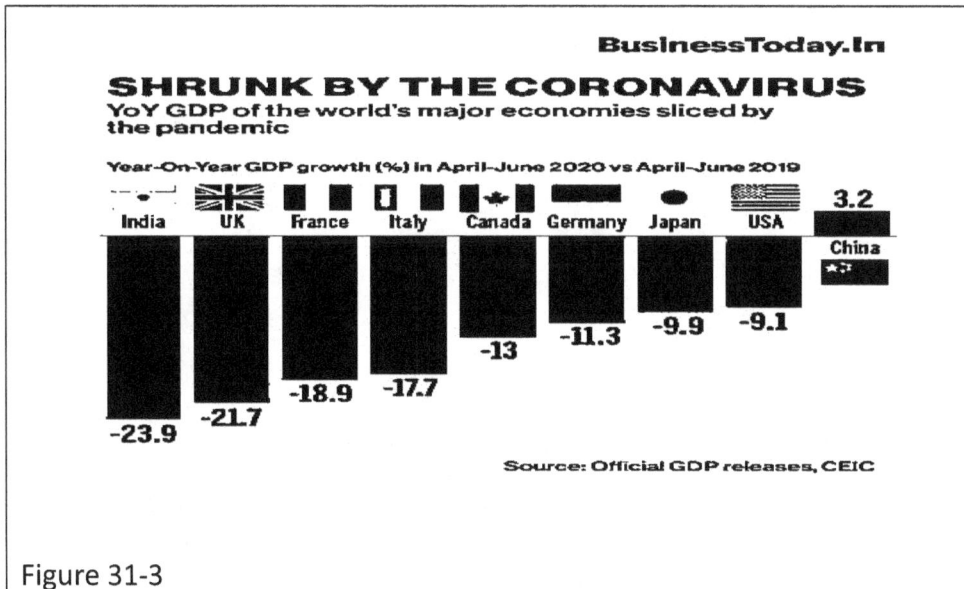

Figure 31-3

At the end of the second quarter of 2020, India had experienced a 23.9% drop in its GDP, while the US had a 9.1% reduction. China, however, had a 3.2% increase in GDP. Most other countries had a 10-20% drop in their GDPs, and the worst was yet to come. (Figure 31-3)

ECONOMIC STIMULUS PACKAGES FROM AROUND THE WORLD

The COVID-19 pandemic forced most countries to deliver stimulus packages, as most workforces had come to a standstill. The stimulus packages ranged from 2.2% of South Korea's GDP to 21.1% of Japan's GDP. South Korea did not even have a lockdown; people continued to live a normal life, using masks

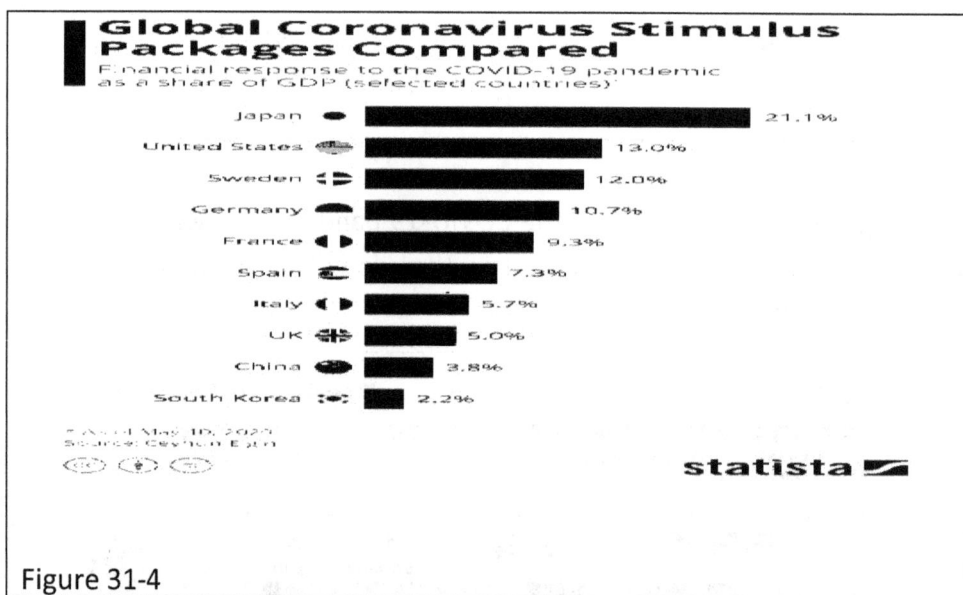

Figure 31-4

The oil demand dried up, dipping oil prices to a record low, made worse by a price war between OPEC and Russia.

THE UNITED STATES COVID-19 STIMULUS PACKAGES

STIMULUS Phase 1: President Trump signed a bill into law on March 6 involving a stimulus package worth 8.3 billion. It provided extra funding for the Centers for Disease Control and Prevention (CDC), Food and Drug Administration (FDA), National Institutes of Health (NIH), the State Department, the Small Business Administration (SBA), and the US Agency for International Development (USAID). The bill also allocated $4 billion for coronavirus tests and $1 billion in loan subsidies for small businesses.

STIMULUS PHASE 2: Pres. Trump signed a $100 billion bill into law providing more financial support for free coronavirus testing, two weeks of paid sick and family leave, increased funding for Medicaid food security programs, and increased unemployment benefits

STIMULUS PHASE 3: Pres. Trump signed a $2 trillion bill on March 27 to provide bailout payments to individuals. It included direct payments of $1200 to individual taxpayers in April/May, costing $500 billion. It also included $300 billion in small business loans, $50 billion bailouts to the airline industry, and $150 billion to hospitality and retail industries.

STIMULUS PHASE 4: Pres. Trump signed four executive orders to address the ongoing coronavirus crisis. It included a $400-per-week payment to unemployed Americans, a payroll tax holiday until the end of 2020, a no-eviction clause for renters, and deferred student loan payments until Dec. 2020. (Figure 31-4)

PAST STIMULUS PACKAGES

In comparison, during the financial crisis of 2008, Pres. Obama authorized $900 billion to help big banks, the auto industry, another $200 billion to rescue mortgage companies.

The 2009 stimulus package, the American Recovery and Reinvestment Act cost roughly $800 billion and funneled money into aid to state and local governments, safety net programs, tax relief, and construction and investment projects.

During the pandemic peak, Brent crude, a benchmark used by Europe and the rest of the world, dipped below $20, to the lowest level seen in 18 years. (Figure 31-5)

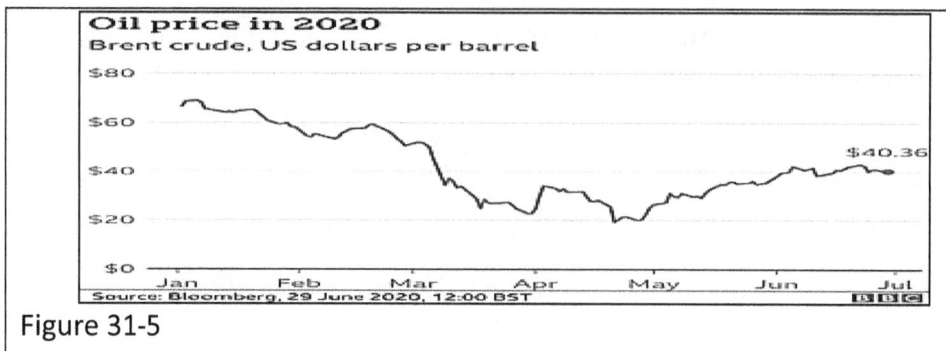

Figure 31-5

AstraZeneca's share price hit record highs. The Drug company was working with the oxford research group to produce and market their coronavirus vaccine. Most of the healthcare stocks did well during this pandemic period. (Figure 31-6)

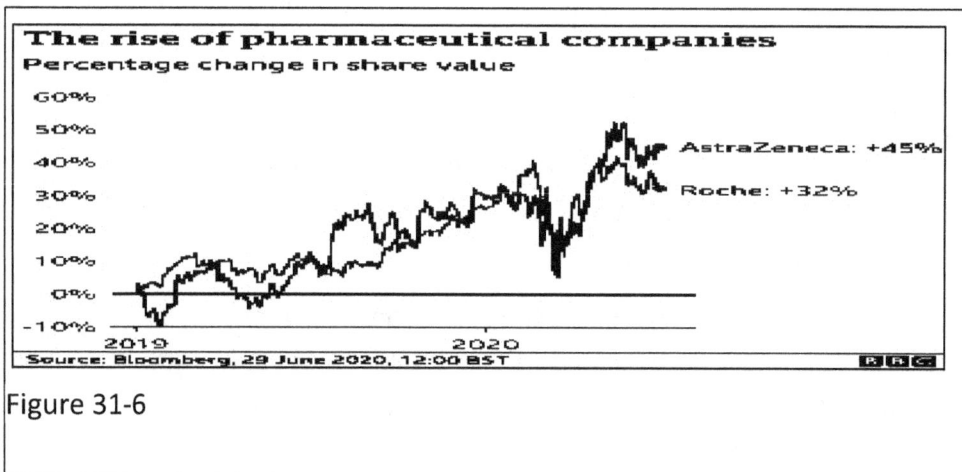

Figure 31-6

Lockdowns resulted in widespread Unemployment, even though the Payment Protection Plan (PPP) in the US provided much liquidity to corporations to keep their employees on the payroll. At the COVID-19 pandemic peak, US unemployment went from a 50-year low of 3.5% to 20% in two months. (Figure 31-7)

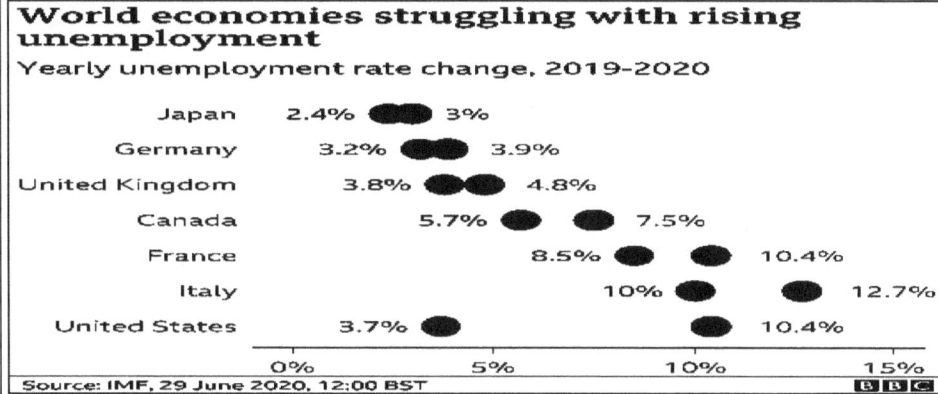

World economies struggling with rising unemployment
Yearly unemployment rate change, 2019-2020

Country	2019	2020
Japan	2.4%	3%
Germany	3.2%	3.9%
United Kingdom	3.8%	4.8%
Canada	5.7%	7.5%
France	8.5%	10.4%
Italy	10%	12.7%
United States	3.7%	10.4%

Source: IMF, 29 June 2020, 12:00 BST

Figure 31-7

During April, a record 20 million Americans filed for unemployment insurance, including 42,000 healthcare workers. By the end of the first quarter of 2020, the US had lost 30 million jobs.

Many countries had to battle between rising coronavirus cases and their economies spiraling down like Niagara Falls.

Places like Florida, Texas, and California, which lifted some of their restrictions too early, ended up with a sudden and unprecedented rise in the

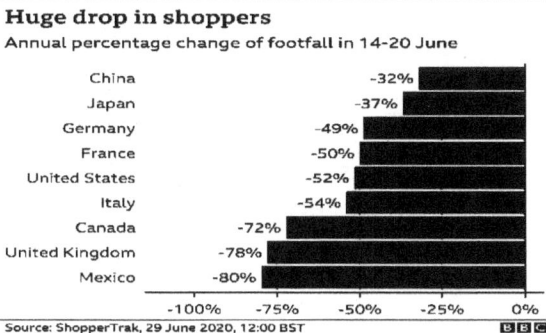

Huge drop in shoppers
Annual percentage change of footfall in 14-20 June

Country	%
China	-32%
Japan	-37%
Germany	-49%
France	-50%
United States	-52%
Italy	-54%
Canada	-72%
United Kingdom	-78%
Mexico	-80%

Source: ShopperTrak, 29 June 2020, 12:00 BST

Figure 31-8

number of new cases, hospital admissions, and increased demand for the ICU beds. This second spike occurred in July and August, even though these states had not fully recovered from the first one. This is an important lesson for future leaders in similar situations.

Economists raised questions about the true economic impact of lockdowns. A University of Copenhagen study showed that Denmark, which instituted social distancing rules, suffered a 29% decline in spending compared to Sweden with no enforced social distancing rules, which had a 25% decline. Most of the economic loss was due to the COVID-19 pandemic itself, not due to the mandated lockdowns. They suggested that social distancing affected supply while the deadly pandemic crushed demand. Humans are highly motivated to be safe, alive, and well. Naturally, most take a conservative approach, which profoundly affects demand. That most countries saw a substantial drop in travel, group meetings, and even family gatherings demonstrated this. (Figure 31-8)

Hence, the fear drove the demand down, while lockdown slowed the supply, and the combination was something most countries had trouble handling.

However, during the recovery phase, the states that instituted self-mitigation and social distancing experienced a robust economic recovery after the pandemic compared to those who had no strict rules.

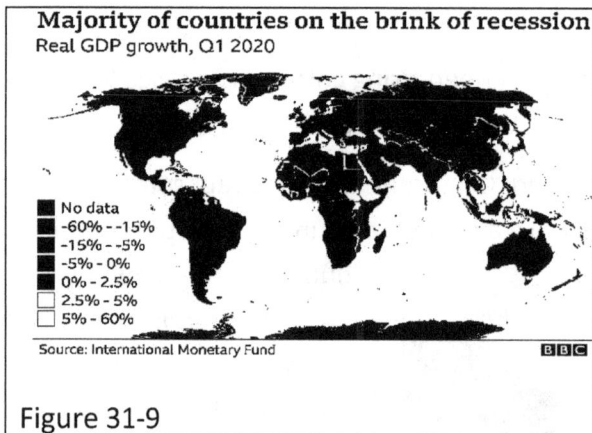

Figure 31-9

This was reflected in the second resurgence wave in the Southern US in Florida, Texas, and Arizona, where people disregarded social distancing after lockdowns were lifted.

The most important lesson is to continue self-mitigation rules and social distancing measures until a vaccine is available, even when lockdowns are no longer mandated.

CHINA: The COVID-19 pandemic coincided with Chinese New Year, which attracted people from all over the world to mainland China. China instructed workers in 31 provinces to stop working until February 10. Those regions accounted for 80% of their GDP and 90% of exports to other countries. China closed schools until March. Its manufacturing also took a tremendous hit.

Most automobile and airline manufacturers had to stop productions from time to time. (Figure 31-9)

Over 200 million students were affected when China closed its schools in Feb. The world bank predicted China's growth that year would be 0.1-2.3%-- the lowest in decades.

HONG KONG: Ongoing protests and unrest had already plagued Hong Kong before the coronavirus epidemic, which made the downward spiral in their economy even worse. There were widespread protests to prevent people from mainland China from entering and spreading the virus.

INDIA: In India, the lockdown especially strangled migrant workers to near starvation with no shelter and no place to go. India suffered the worst economic loss in the world from the COVID-19 pandemic, all while it was still struggling during its first wave.

JAPAN: Japan suffered a major drop in tourism and a shortage of face masks. Japan canceled major sporting events and postponed the 2020 Tokyo Summer Olympics to 2021.

FRANCE: The bank of France declared that the French economy shrunk 6% in the first quarter of 2020. The country was in a recession. Similarly, the Deutsche Bank reported that Germany was in a recession.

CANADA: On 4 March, Canadian Prime Minister Justin Trudeau announced that their government had allocated CA$1 billion for the COVID-19 response.

Ref:

https://www.bbc.com/news/business-51706225

https://www.statista.com/topics/6139/covid-19-impact-on-the-global-economy/

https://arxiv.org/abs/2005.04630...

https://papers.ssrn.com/sol3/papers.cfm?abstract_id=3561560

https://pubs.rsna.org/doi/full/10.1148/radiol.2020200843

https://www.nejm.org/doi/full/10.1056/NEJMc2009787

https://www.ajkd.org/article/S0272-6386(20)30618-1/fulltext

https://jamanetwork.com/.../jamacardiology/fullarticle/2763846

https://www.cambridge.org/.../204BD93C135EC727FAEFC62E3BE72C3B

https://bfi.uchicago.edu/working-paper/2020-26/

https://www.cam.ac.uk/.../economic-damage-could-be-worse-with...

https://link.springer.com/article/10.1007/s10654-020-00649-w

32 EDUCATION CHALLENGES AT THE COLLEGE LEVEL

As the relentless rise in COVID-19 cases, increased occupancy of hospital beds, and increased demand for ICU beds occupied the minds of business people and political leaders, universities faced another set of challenges with how to proceed with the fall semester.

More than 50% of states had seen a steep rise in new cases in the second wave, often as much as 100% or more compared to that of their initial wave. Some increase in the numbers is attributable to people ignoring self-mitigation guidelines and gathering at beaches and bars and taking part in protests and riots. There also was an increase in the number of younger people being admitted to the hospitals.

Universities faced far-reaching challenges to keep the students safe while feeding their young minds and transforming them into productive graduates. The first task was how to maintain social distance and provide the required engaging learning atmosphere.

STUDENTS

Some students were unsure if they were going to have any on-campus classes in the fall. For new students, being in an unfamiliar environment, maintaining social distance, wearing a mask all the time while also and learning presented a unique challenge, especially if they would also be sharing a small dorm space with 2-4 other students. Meeting other students and having an active college lifestyle in the COVID-19 era is a challenge, as students cannot have close interactions in class or meet in groups in libraries. Many students faced the possibility of taking all their classes online from their dorms, potentially drastically limiting their interaction with others and exacerbating cabin-fever. Traveling between their residence and the university presented an additional challenge due to air travel restrictions.

International students faced new visa requirements and restrictions. Their ability to enroll in classes depended on whether there was an international flight service available from their country. Officials restricted students entering from certain countries due to the coronavirus pandemic.

How many fresh graduates would have preferred to do the first semester of college life online? I didn't think so!

PARENTS

As a parent who had three kids in college at the same time over five years, I might say I never attended the University of Texas, but most of my money went to UT at Austin.

Whether students stayed at home for an online college or attended on-site college, enrolled in a hybrid of those options, colleges charged the same or more in tuition dollars. A survey of 25 colleges disclosed tuitions ranging from $10,000 to $39,000 per year, not counting the food, housing, clothes, books, and air travel.

Parents worried about whether their grown-up adults were going to maintain the social etiquette demanded of them during the COVID-19 crisis. FaceTime policing would not work.

UNIVERSITIES

Universities are mega-institutions with multimillion-dollar budgets. In the COVID-19 era, they had to focus on admissions, students, classrooms, sports, extracurricular activities, ball games, faculty, their image, all alongside student health and welfare.

Universities were keen on providing an on-campus experience, which was their signature attraction; however, physical constraints posed a challenge. If they did not offer an on-campus experience, students might switch to choose a university that did. So, they would be in competition for the same student body. Everything depended on how universities marketed their programs.

Many universities moved classes online early in the pandemic in the Spring semester. With no clear end to the pandemic at the beginning of the Fall semester, it was clear that coronavirus case numbers could dramatically increase in colleges providing a full on-campus experience.

The genuine dilemma for most universities was to convince their faculty that it was in their best interest to invest their own time and resources in learning the nuances of presenting online classes effectively.

Because of intense competition for a comprehensive experience, some lesser-known institutions might go under if they cannot attract enough students to their programs. The top 25 colleges and universities might get their share of students at premium tuitions, as they always claimed that they had 85%–87% rejection rates during the past years. Attracting the students was one thing but providing the desired educational and cultural experience in an isolated manner was a much bigger challenge.

How were they going to handle the cafeteria, sporting events, or group activities? A virtual sporting event or a gathering was a dream, and sports and sporting events generated millions of dollars for the university through games, sponsorships, and donations.

Some smart universities tried to team up with tech giants like Apple, Google, or Amazon to provide new work-related study programs sponsored by those companies.

Universities also had to play the role of a healthcare provider as they would invariably face COVID-19 infections on campus. This required the development of a comprehensive program for detection, quarantine, and contact tracing if a coronavirus outbreak were to happen on one or their campuses. Several campus healthcare workers and caseworkers were required to deal with outbreaks.

Universities had to have central air conditioning in dorms and campus buildings checked to minimize the coronavirus spread. When New York City was on lockdown, several new cases arose from multi-story crowded apartment buildings, indicating that spread between buildings can occur with improperly maintained air systems.

Some universities redesigned the semester to cut it short and let the students go home after thanksgiving.

Many universities reopened their research facilities in June or July. Some planned to bring students back in phases. Many contemplated having online instruction after thanksgiving, including exams.

Some colleges planned to have half of their students on campus and half learning remotely.

FACULTY

The greatest responsibility fell on the shoulders of the faculty, who were under undue pressure to deliver whatever the universities promised. Faculty had to please the administrators and amuse the new students with their dexterity in not only teaching the subjects but also in managing students and the technology in an online class. If the technology hit a snag, they would not have a second chance. They had to watch a lot of training videos on the pros and cons of what to do when they delivered an online class. The faculty could not blame the university for their ten-year-old laptop.

Some faculty members had to take a pay cut or even retire. Elderly faculty members were more concerned about exposure to the coronavirus than delivering eloquent lectures. They were never too old to learn new things, according to the administration.

Some universities froze faculty salaries or ceased contributing to retirement funds as they were expecting a steep revenue shortage.

Some universities considered restricting in-person instruction to only those that needed it. Many universities were talking about moving exams online.

Most universities required everyone to wear masks while on campus and in classrooms. Many had plans to provide coronavirus testing on the campus.

Safety came first, and while most students have returned to their universities for the fall semester, what follows will be determined by any change in case numbers and infection rates. Plans can completely change overnight as the coronavirus whimsy shifts the tides.

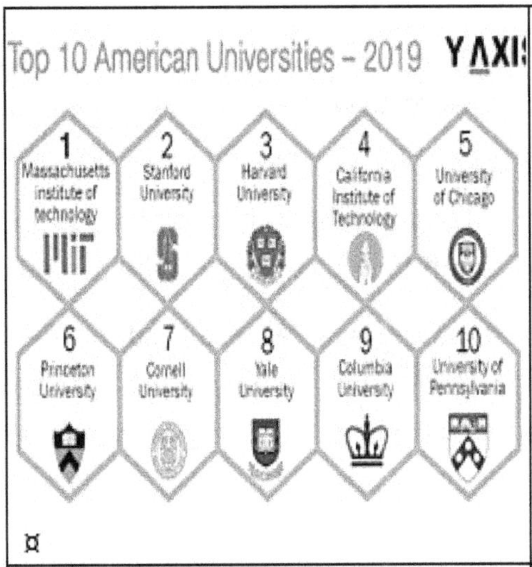
Top 10 American Universities – 2019
1
Massachusetts Institute of technology
2
Stanford University
3
Harvard University
4
California Institute of Technology
5
University of Chicago
6
Princeton University
7
Cornell University
8
Yale University
9
Columbia University
10
University of Pennsylvania

BEST
COLLEGES
U.S.News
& WORLD REPORT
RANKINGS

NEVADA

BOSTON
COLLEGE
OHIO
STATE
JOHNS HOPKINS
DREW
NYU
FORDHAM UNIVERSITY
LAFAYETTE
THE UNIVERSITY OF
CHICAGO
Tulane
Tufts
The
UNIVERSITY
of VERMONT
Bates
Universities in USA

Universities

University

33 SHOULD SCHOOLS RESUME IN-PERSON CLASSES IN FALL 2020?

As National Conventions were shifting to an online format and the Southern States such as Florida, California, and Texas experienced record increases in COVID-19 cases, surpassing infection rates that flooded New York hospitals in early March, a spirited debate emerged on the return to in-person teaching for grades K1-12. It was not possible to predict how children would respond to a classroom experience amid a rapidly changing pandemic.

The long-term effects of isolation in adults, such as income and job loss, an increase in emotional and psychological distress, and an increase in domestic violence, have been well-documented during COVID-19 and past pandemics. Anxiety, depression, and insomnia have increased among quarantined populations, with a knock-on effect on children as parents adapt to new, unexpected roles in homeschooling and round-the-clock parenting while juggling fragile jobs or facing unemployment and other financial concerns.

THE IMPORTANCE OF IN-PERSON EDUCATION

Teachers and schools provide essential services that shape children's maturity, creativity, and interpersonal skills to prepare them for a prosperous future. In-person classroom education in schools enables parents, guardians, and caregivers to work, and schools mitigate health disparities by providing meals and access to physical, behavioral, and mental health services. Thus, schools play a critical role in supporting children beyond the academic level.

School closures place a significant burden on parents who may be unable to provide adequate supervision and educational support at home due to work commitments, and many schools lack resources for a full, effective transition to online learning. Children with disabilities may not have access to support required for their Individualized Education Programs when classrooms are online. Thus, distance learning may increase the disparities in educational outcomes that existed pre-COVID-19 across income levels, racial, and ethnic groups, potentially worsening the educational, health, and economic well-being of the community as well as within individual families.

In-person interaction is critical for the development of language, language, communication, emotional, social, and interpersonal skills. Schools provide a social support network that builds coping skills for challenging situations, goal setting, interpersonal relations, and fostering positive relationships. In addition, the traumatic nature of the COVID-19 pandemic may be partially mitigated by the structure and routine of school life, including physical activities, which are beneficial for reducing anxiety and depression. Coupled with behavioral assessment and counseling services for critical psychological and behavioral needs, schools boost student mental wellbeing through support and services, which may be inaccessible in an online format.

There were many compelling reasons for opening schools in the fall of 2020. However, it was essential to weigh the risks and benefits and approach a return to in-person education with a clear, safety-focused, comprehensive plan.

COVID-19 INFECTION IN CHILDREN

COVID-19 symptoms in children may include fever, headache, sore throat, cough, fatigue, nausea/vomiting, and diarrhea. Like adults with COVID-19, many children could be asymptomatic. To date, children experience less severe COVID-19 symptoms in comparison with adults—

fewer than 5% of infected individuals are children, and in patients under 17 years of age, the mortality rate is less than 0.1%[8]. Infection rates in children, like observations in adults, vary by race. Hispanic/Latinx children are testing positive at a higher rate than non-Hispanic white children. Risk increases with age, and students in the fifth grade and beyond may be more susceptible to COVID-19 than those under ten years of age.

Children who have pre-existing medical conditions may have a significantly increased risk for severe illness from COVID-19. Students with intellectual and developmental disabilities are disproportionately affected by additional conditions such as respiratory, circulatory, and metabolic diseases that are known to increase mortality rates in adults infected with COVID-19. Severe complications are potentially fatal and may require hospitalization, intensive care, and/or ventilation. In rare cases, children may develop a serious complication termed multisystem inflammatory syndrome (MIS-C) after COVID-19 infection, which can be fatal.

ADAPTING IN-PERSON EDUCATION IN A PANDEMIC SETTING

The CDC has offered specific recommendations to school administrators for maintaining health and safety during a return to in-person teaching. Under these guidelines, school staff and community members are encouraged to practice preventive behaviors such as social distancing in all areas and canceling activities in which social distancing is not possible, wearing masks, routine cleaning and disinfecting of all classrooms and other common areas in the school, use of outdoor facilities whenever possible, and regular handwashing. The CDC recommended incorporating the education of staff and students on safe COVID-19 practices into the academic curriculum and extracurricular activities along with altering the day-to-day functioning of schools with safety in mind.

One option presented by the CDC was to use a "cohorting" approach in which students are kept in a single group with one set of teachers throughout the school week. This is challenging in middle and high school settings due to large class sizes and the range of school subjects taught but may be appropriate for younger students. Other strategies included staggering schedules and separating students by grade to achieve small groups. The rationale behind these small-group strategies was that if a student, teacher, or staff member tests positive for COVID-19, only the members of that small group would need to be tested and quarantine, limiting overall transmission and absences. Communities may also be equipped to aid the mitigation of transmission, providing access to large facilities such as church halls where groups of students can gather while maintaining the necessary physical distance.

MITIGATING COVID-19 TRANSMISSION DURING IN-PERSON SCHOOLING

Many countries outside of the US resumed in-person classroom teaching with mixed results. Lessons from these countries provided valuable insight into preparations for US schools; for example, China, Denmark, Norway, Singapore, and Taiwan all required temperature checks before entering school facilities. Community cases increased in Denmark after schools and childcare centers reopened; however, cases have since steadily among children aged 1-19 years. Taiwan instituted mandatory face masks and entrusted schools to develop reopening and temporary closure methods in accordance with local infection rates, and many schools reduced class sizes, increased physical distancing, and implemented cohorting.

Israel experienced a surge in cases with 503 students and 167 staff members testing positive in schools after reopening and relaxing of social distancing measures, resulting in a temporary closure of 355 of its ~5000 schools, though mitigation methods used were unclear. Here in the US, over 1300 positive cases were reported in Texas childcare centers with twice as many staff members affected as students. This corroborates other findings that suggest that children may less susceptible to COVID-19 infection than adults and that COVID-19 infections in children typically result from transmission from other COVID-19 positive family members. In studies following school reopening in France, Australia, and Ireland, it was determined that children might be less likely to transmit the virus among each other in comparison with household contacts.

International data indicate that new cases will arise with the community and schools despite careful coordination, planning, and preparation. For this reason, it is critical for schools to keep up to date with local resources on COVID-19 recommendations and responses and develop a proactive plan for symptom monitoring, contact tracing, and isolation when a student or staff member tests positive for COVID-19. Maintaining an up-to-date record of absences allows for efficient tracking of the impact of COVID-19 on student and staff absences beyond typical absence rates.

Regular communication with local officials regarding daily transmission rates and the impact of this on students and parents is critical, and education must be passed on to students to mitigate the spread of infection, such as explaining methods to reduce the risk of contracting the virus from a parent or other household member. Consistent communication between the school administration, staff, children, and their parents is also recommended by the CDC to ensure everyone is informed of current conditions and make sure the necessary steps are taken to mitigate outbreaks. Families and schools must be mutually transparent about transmission and outbreaks within the community while maintaining confidentiality per HIPAA.

The CDC recommended that schools have systems in place to support the continuity of learning for students who must isolate or quarantine at home following COVID-19 infection or exposure. Family members or anyone in close contact with a person who has tested positive or is symptomatic for COVID-19 were advised to get tested and stay home until receiving a negative result or stay home and monitor for symptoms, as per CDC guidelines. Thus, schools must educate members on when to seek medical help, when to isolate and quarantine, and how to transition back to the in-person environment after the end of the isolation or quarantine period, and schools must plan to provide access to online learning, school meals, and other critical services to all students unable to attend in-person.

In the event of a substantial transmission within the school or community, schools must work with local health officials to determine whether to maintain school operations and, if temporary closure is necessary, how long that closure should be. One positive case within a school may not warrant closure; however, outbreaks originating in a school setting might require a temporary return to distance learning.

OPPONENTS' POINT OF VIEW

Due to the steep rise in COVID-19 cases in Southern states with a record number of hospital admissions and overwhelmed ICUs along with the prospect of a second wave, many individuals considered in-person schooling a significant risk for children at this time. The spike in COVID-19 cases in South Korea following school reopening and the subsequent necessity of further school shutdowns supported this position.

Despite measures put in place to enhance safety, children often find it difficult to wear face coverings for extended periods and touch and continuously adjust these coverings, rendering the masks counterproductive and a potential source of infection. Children are also at risk of contracting the virus at school and, in turn, infecting other members of the family who may be more vulnerable. However, children are less likely to spread the virus, and most infections in children are community-acquired.

Additionally, there were disparities in opinion on how to approach school funding in a distance-learning setting. Schools receive billions of dollars each year to provide a balanced educational experience to the students from K1-12 grade. Some felt that money allocated to education should go to the parents if children were not receiving in-person education to allow parents to invest in providing quality education to their children.

The counterargument was that taking away funds from inner-city schools, which were already strapped for money, would lead to deterioration of the schools to the point that they may not be safe learning environments in the future. Schools needed money to invest in virtual technology in a distance learning setting and to provide low-income students with the hardware, software, and training required to use virtual learning platforms effectively. These efforts required more funding, not less.

CONCLUSION

Overall, evidence from schools worldwide indicated that in-person education was likely to be safe in states where community spread was minimal or declining; however, it would be wise to proceed with caution in areas in which case numbers are increasing rapidly.

Ref:

https://www.cdc.gov/coronavirus/2019-ncov/community/schools-childcare/prepare-safe-return.html

https://www.sciencemag.org/news/2020/07/school-openings-across-globe-suggest-ways-keep-coronavirus-bay-despite-outbreakss

Sherr, L., Roberts, K.J. & Gandhi, N. Child violence experiences in institutionalized/orphanage care. Psychol Health Med 22, 31-57 (2017).

Bradbury-Jones, C. & Isham, L. The pandemic paradox: The consequences of COVID-19 on domestic violence. J Clin Nurs 29, 2047-2049 (2020).

Roesch, E., Amin, A., Gupta, J. & Garcia-Moreno, C. Violence against women during covid-19 pandemic restrictions. BMJ 369, m1712 (2020).

Lyons, D. et al. Fallout from the Covid-19 pandemic - should we prepare for a tsunami of post-viral depression? Ir J Psychol Med, 1-10 (2020).

Caley, P., Philp, D.J. & McCracken, K. Quantifying social distancing arising from pandemic influenza. J R Soc Interface 5, 631-9 (2008).

Tang, F. et al. COVID-19 related depression and anxiety among quarantined respondents. Psychol Health, 1-15 (2020).

Altena, E. et al. Dealing with sleep problems during home confinement due to the COVID-19 outbreak: Practical recommendations from a task force of the European CBT-I Academy. J Sleep Res, e13052 (2020).

National Center for Health Statistics, C.f.D.C.a.P. Preparing for a Safe Return to School. (2020).

Loades, M.E. et al. Rapid Systematic Review: The Impact of Social Isolation and Loneliness on the Mental Health of Children and Adolescents in the Context of COVID-19. J Am Acad Child Adolesc Psychiatry (2020).

Danis, K. et al. Cluster of Coronavirus Disease 2019 (COVID-19) in the French Alps, February 2020. Clin Infect Dis 71, 825-832 (2020).

(NCIRS)., N.C.f.I.R.a.S. COVID-19 in schools – the experience in NSW. . (NCIRS, Westmead, 2020).

Heavey, L., Casey, G., Kelly, C., Kelly, D. & McDarby, G. No evidence of secondary transmission of COVID-19 from children attending school in Ireland, 2020. Euro Surveill 25(2020).

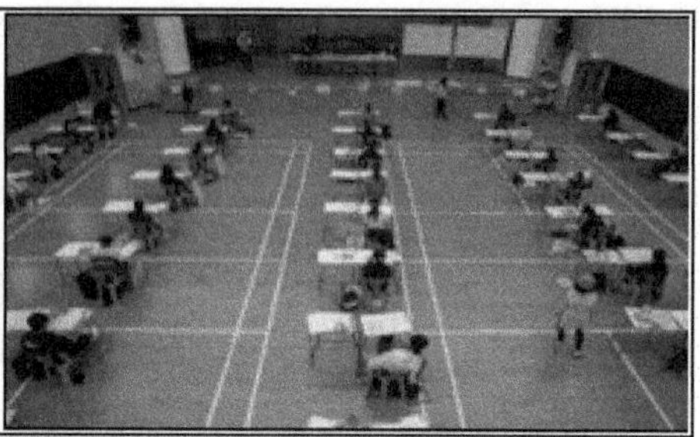

SCHOOL

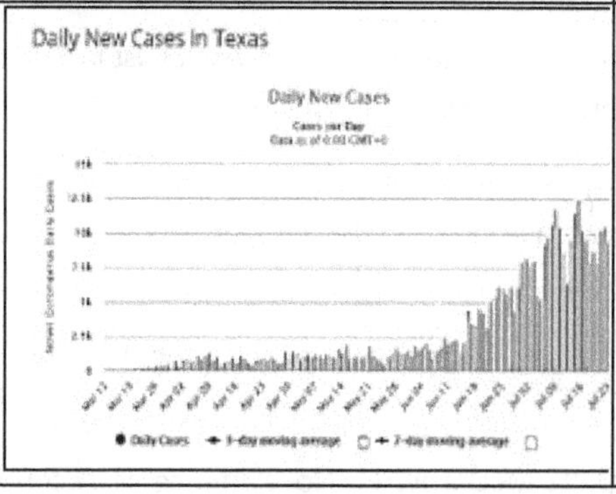
Daily New Cases in Texas
Daily New Cases

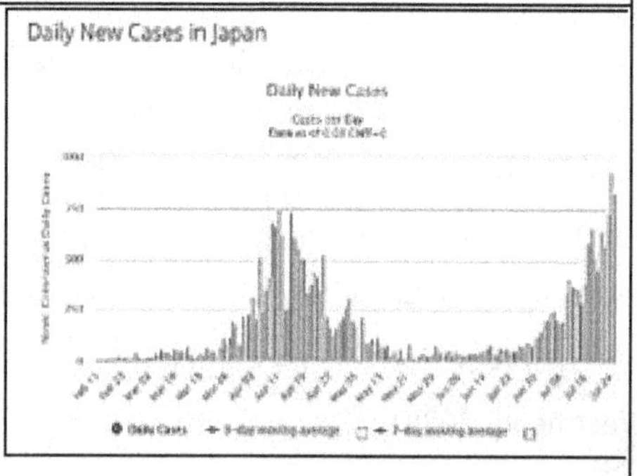
Daily New Cases in Japan
Daily New Cases

34 SECOND WAVE OR SURGE AROUND THE WORLD

DIFFERENT CAUSES AND DIFFERENT RESPONSES

When the first wave of COVID-19 pandemic lost pace in May and June in many parts of the world, people thought the summer heat would choke the virus and drastically reduce the number of new cases.

Contrary to this wishful belief, the number of new cases spiked for the second time with more vengeance in the southern US, followed by a similar second wave or surge in other countries around the world. Even in countries where cases had almost disappeared, such as New Zealand, Australia, and Japan, COVID-19 cases again began to rise. The second wave was bigger than the first one in Australia, Japan, Hong Kong, the UK, Sweden, France, and South Korea.

WHAT TRIGGERED THE SECOND WAVE OR SURGE?

The reasons were many and varied, depending on the countries. Many states and countries ended a total lockdown. Some felt they ended the lockdown prematurely. That was the concern in Texas, Florida, and California, which experienced an early and unsuspected second wave. (Figure 34-1)

Many major cities in the US faced new challenges in the form of nationwide unrest, protests, and large gatherings. Thousands of people ignored the social distancing and self-mitigation rules as the momentum spread from one city to the next.

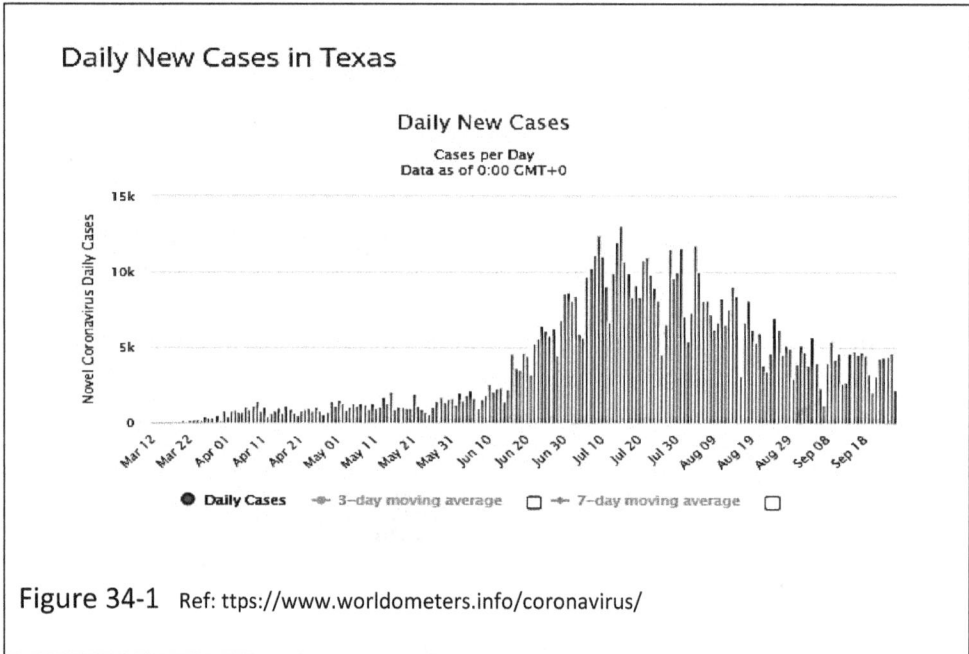

Figure 34-1   Ref: ttps://www.worldometers.info/coronavirus/

It was not clear if the large gatherings had contributed to the increased percentage of people testing positive and increased hospitalizations. Around that time, 40% of admissions in Houston were people aged 20 -40, indicating that the virus had shifted from affecting the older individuals to the younger population.

The percentage of people testing positive in Texas went up from 4% to as high as 21% in Houston during its peak at the beginning of September 2020. The opening of beaches, bars, and restaurants contributed to the sudden rise in new cases. There was also a silent gathering of thousands of people in various parts of the country, defying the lockdown and expressing their freedom of speech. Most of them ignored mask mandates and social distancing guidelines.

This occurred in response to the weeks-long strict lockdowns where people many people were unemployed, restless, and facing financial hardships.

On the public health side, coronavirus tests were taking too long to process, with results delayed over eight days in some cases. States lacked financial and human resources and trained personnel required for contact tracing.

EFFECT ON THE HEALTHCARE SYSTEM

New cases and hospital admissions almost doubled and tripled over a few weeks in Texas, Florida, and California, taking a major toll on healthcare resources. The ICU occupancy far exceeded bed capacity, and inpatient numbers climbed swiftly.

Increasing testing came with a price tag. More active infections were discovered, causing total daily new cases to skyrocket. A major concern arose when the percentage of people testing positive climbed from 4.0% to as high as 21% (San Antonio—July). That meant that one out of every five people who were tested had COVID-19. (Figure 34-2)

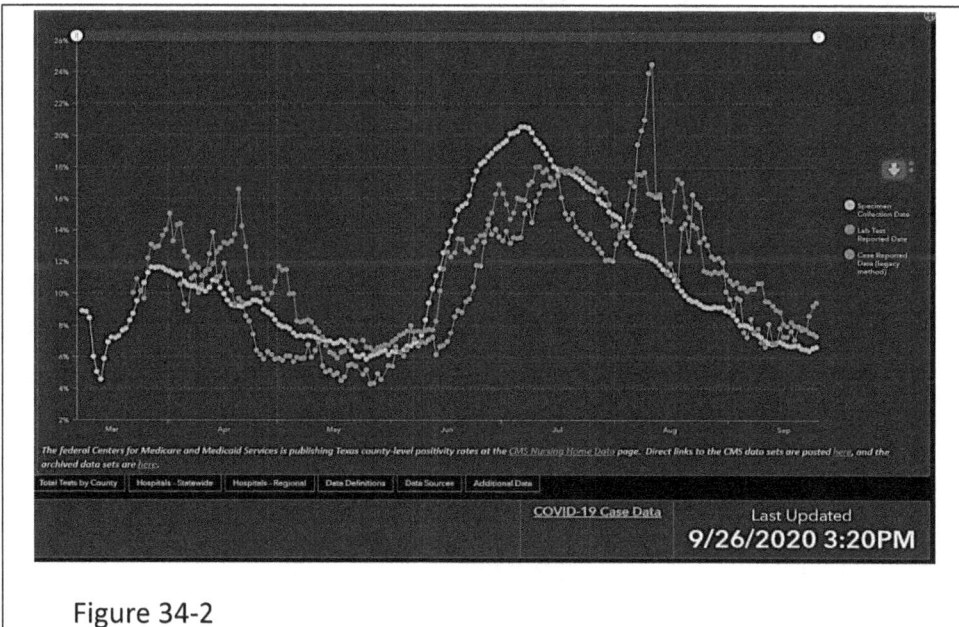

Figure 34-2

This signaled two things. First, that social unrest, the gathering of people in many forms, and the disregard for self-mitigation guidelines took a toll on the state numbers. Second, that the pandemic was getting out of control again, and mitigation methods or total lockdown might be required again for an extended time.

COVID-19 cases did not spike, however, in the 3.5 million people who had been working since the pandemic began in the service and transportation industries.

Island countries like New Zealand and Australia with much smaller populations were highly successful in controlling their borders and the COVID-19 pandemic. However, they could not stay isolated forever.

When they let their citizens return from foreign countries in July and August, they saw a fresh surge of new cases. The stories were similar in Japan, Hong Kong, South Korea, and Germany, to name a few.

Resumption of air travel was another major concern as most countries could not rely on other countries to follow the same mitigation levels as they had in place in their countries. New York instituted a voluntary 14-day quarantine for those visiting the state from other states with high infection rates, such as Texas, Florida, or California.

It is important to remember that a viral pandemic can make a second and perhaps a third round before it fades away into the sunset.

SECOND ROUND MEASURES

Most people were restless and drained from the first lockdown and not prepared for a second one. The economic toll from the first lockdown had left many people with no jobs and income. Another total lockdown was not a practical or viable option. Additionally, public opinion was deeply divided and polarized. It became difficult to decipher truth from frustration and anger from both sides.

The focus shifted to limiting the gathering of large groups of people. So, states like Florida, Texas, and California banned attending beaches, bars, and large gatherings. Italy closed all bars until Sept. 7.

Testing and isolating active cases became a central theme. That was a dream in most countries where it would require more than 100,000 caseworkers to trace a small cluster of the population.

COVID-19 did not differ from past viral pandemics in many respects. It ran its destructive course throughout the globe, with more than 33 million cases and more than 1,000,000 deaths (Sept. 27).

People are people, and some went against the grain. In the process, many became victims of COVID-19.

A few options left for most countries included continuing self-mitigation principles, letting people return to work, and reducing or minimizing gatherings, all with the hope for a breakthrough vaccine that might provide relief.

The major fallouts were social gatherings, parties, bars, and night clubs.

Many EU countries restricted travel from places with a severe surge in new cases, such as Spain, Austria, and Croatia.

The fourteen-day self-quarantine for the travelers arriving from other countries limited travel for businesspeople as they needed to conduct meetings face-to-face.

TEXAS

Texas did well during March and April, and even though their numbers had not come down to zero during May or June, the lockdown was lifted at the end of June. In only a couple of weeks, cases jumped from 2000 per day to more than 10,000. It was only toward the middle of July when new cases began to decline again.

Yet, on August 22, Texas still recorded 7000 new cases per day. Texas did not implement a second lockdown but restricted large group gatherings. There was no restriction on domestic traveling, and, unlike New York, Texas did not require people to quarantine if they came from another state. Florida implemented similar measures, including re-closing beaches, which was unpopular among the young and the restless. By Sept 27, new cases in Texas had gradually declined to 4000 per day. (Figure 34-3)

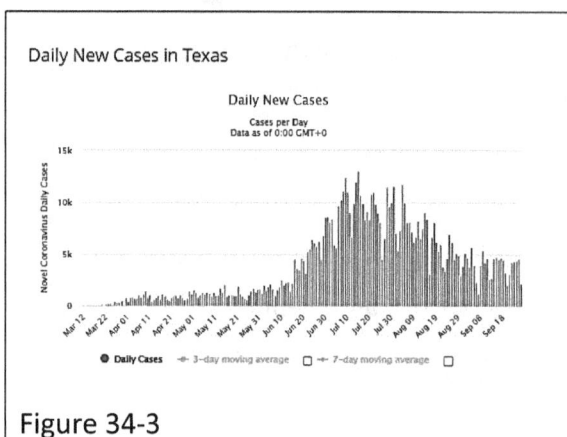

Figure 34-3

During the second surge, 24% of tested individuals were positive for the virus. That signaled good and bad news. It was bad as it overwhelmed that healthcare system; however, with so many people being exposed to the coronavirus, the number of people with temporary immunity was higher, thus possibly reducing viral spread in the future.

CALIFORNIA & FLORIDA

The California and Florida graphs very much mirrored the Texas graph. Their case numbers jumped beyond 10,000 new cases per day in late July and early August. Governor Newsom had to ban the gathering of large groups of people. The reasons were very much identical. (Figure 34-4)

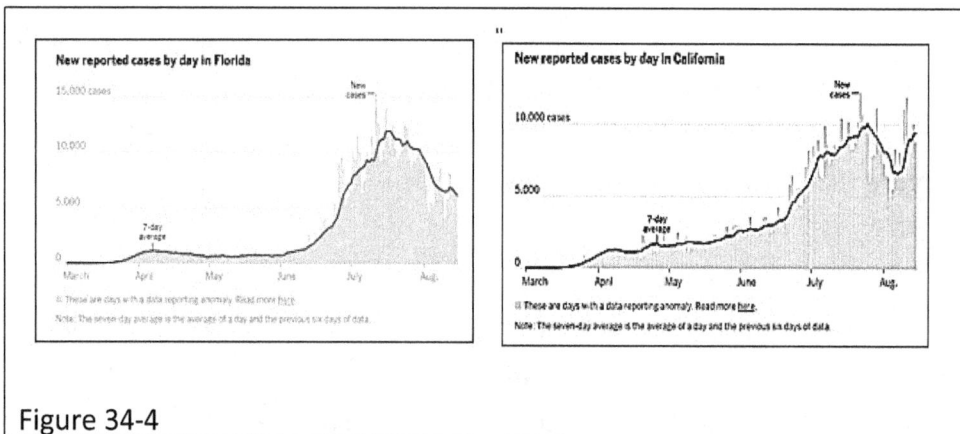

Figure 34-4

NEW ZEALAND

New Zealand was at a loss to explain how the coronavirus had returned after more than 100 days. The hotel quarantine program had failed. Twenty imported cases in the hotel quarantine sites were unexpected, and the result did not go well with the administration. The military was put in charge of overseeing the hotel quarantine program. (Figure 34-5)

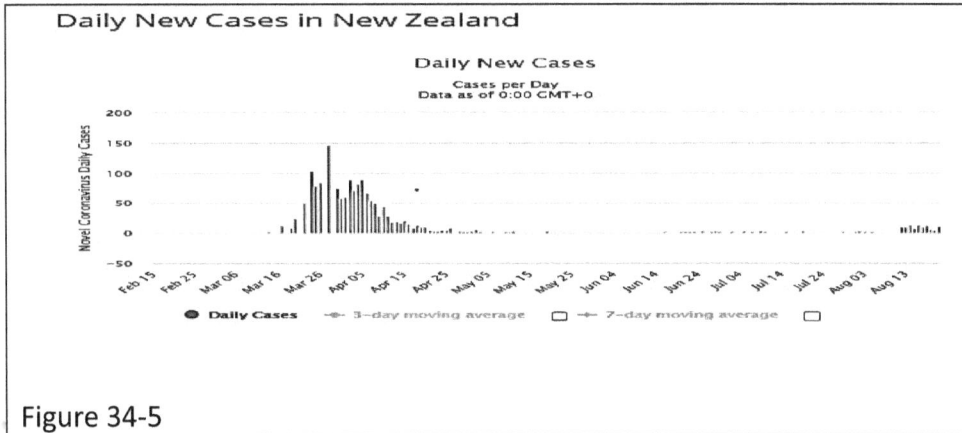

Figure 34-5

Regular testing at the border also failed to meet expectations. Speculations circulated that COVID-19 could have traveled in with frozen foods from abroad. There was no evidence the virus was "brewing in the community."

At the beginning of August, New Zealand had 12 new cases. Auckland, a city of 1.7 million people, decided to lock down until August 26. On August 18, they had 56 cases.

Another source was international flights. The cabin crew was exempt from the quarantine imposed on travelers. Some asymptomatic carriers might have passed on the virus to other people at the airport and hotels. COVID-19 resurgence delayed New Zealand's general elections for four weeks.

AUSTRALIA

By August 4, Melbourne, Australia became the center of a spiraling second wave, forcing a reintroduction of lockdown. Melbourne, the capital of food and culture with a population of 5 million, implicated a stage-3 lockdown in early July. Face masks became mandatory on July 22.

Fines of 4,957 Australian dollars were imposed for violating the quarantine. Meatpacking and construction industries were forced to scale back their work hours and workforce. (Figure 34-6)

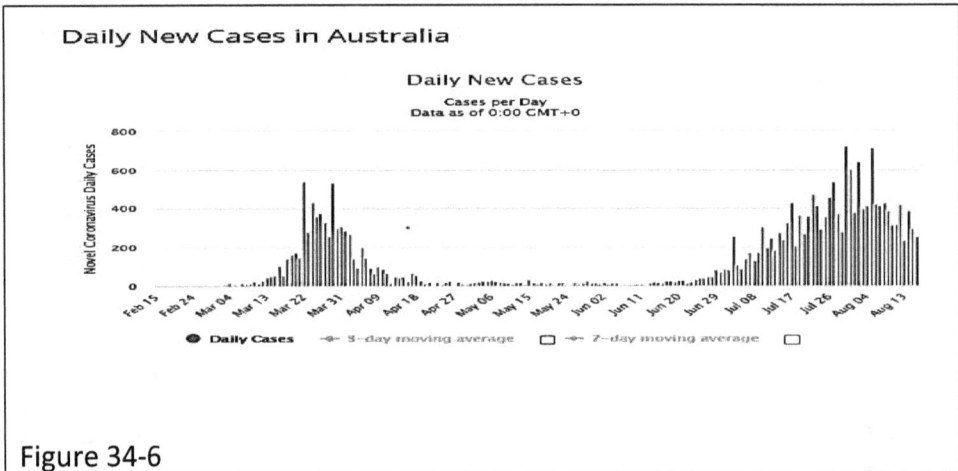

Figure 34-6

It was speculated that the virus could be suppressed if over 70% of the population followed social distancing guidelines and public health rules.

Australia signed an agreement with the UK through the Oxford Vaccine program to acquire 25 million vaccine doses and were contemplating making vaccination mandatory for the entire population.

JAPAN

Japan's earlier success in suppressing the coronavirus outbreak was shattered by a raging flare-up of infections nationwide, more severe than the initial one in March and April. (Figure 34-7)

Tokyo became a major hotspot, with an upward trend in July signaling an impending second wave. Their case counts went from fewer than 100 per day in June to over 1000 per day in July. Toward the end of July, cases spiked to 1500 per day. Tokyo was not sure if it could weather the latest outbreak.

The government delicately tried to balance its economic consumption and its reputation as a safe host for the 2021 Olympic Games in Tokyo.

Over 50% of new cases were untraceable. Alarming clusters of infection were noted in households, sporting groups, and dormitories, along with bars and karaoke venues, which were more expected to be sources of infection. As in the US, Japan's second wave affected a record number of people in their

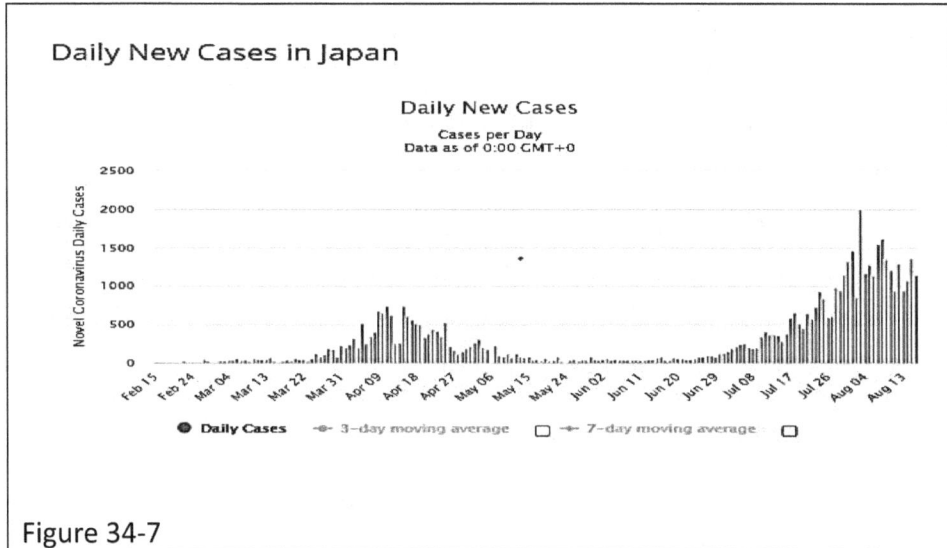

Figure 34-7

twenties and thirties who could expose vulnerable older people. The lockdown corona fatigue may have led to complacency, especially among the younger groups.

The central government tried to avoid a second lockdown. It worried about the economic impact it had experienced from the first strict lockdown. However, the Tokyo governor felt he might have to declare a state of emergency if the trend continued at an accelerated pace. Karaoke venues and bars closed by 10 pm until the end of August. Cash bonuses were offered as an incentive for businesses to comply with regulations.

Okinawa Island declared a two-week state of emergency to prevent the collapse of its health care system.

Tokyo set aside 2,400 beds for COVID-19 patients and 100 beds for critical patients. Over 500 active COVID-19 patients were under home quarantine.

People criticized the government for launching a local tourism campaign on July 22, during the peak of the pandemic.

Japan launched a contact tracing app on June 19. Despite 9.5 million downloads, only 76 positive cases registered. People feared being tracked by the government and exposing their health records.

The local and international media praised Japan's low COVID-19 deaths, despite a high population density (6,158 persons per square kilometer), as a success story. After lifting the state of emergency in mid-May, Japan adopted a controversial approach of implementing voluntary self-mitigation steps and limited testing for the virus.

By June, restaurants and bars opened in full force, and the sumo wrestling and baseball resumed. Japan's government announced $2 trillion in stimulus spending, equivalent to 40% of its gross domestic product (GDP).

HONG KONG

The people of Hong Kong were so convinced they had eradicated the coronavirus that they let their guards down and ignored social distancing guidelines. (Figure 34-8)

Some reasons for their second wave were cited as public transportation,

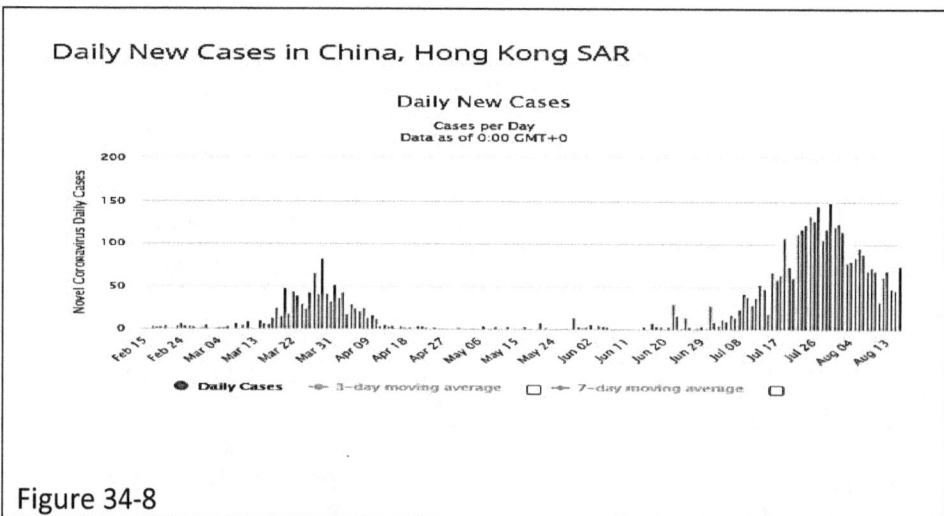

Figure 34-8

where social distancing was impossible, the reopening of schools, and travelers from abroad bringing the virus into the country again.

Experts slammed the Hong Kong government's decision to waive tests and quarantine for over 200,000 arrivals, instead implementing a measure to allow people to self-quarantine with electronic monitoring devices. Social distancing was reimplemented.

"Closed, air-conditioned places had a 19-times higher risk of COVID-19 transmission than outdoor spots. Hong Kong banned daytime dining in restaurants in July, only to reverse their decision the next day, after a public outcry.

Hong Kong recommended that organizations keep workers at home as much as possible and arrange flexible working hours. Hospitals were recommended to test all inpatients who had respiratory symptoms. They stockpiled N95 masks and PPE for healthcare workers.

GERMANY

The German doctors' union warned of a second wave, with the potential to be bigger than the first, due to the public ignoring the social distancing measures.

Germany had one of the best healthcare systems in Europe, with plenty of hospital bed capacity, central health policy initiatives, and a far superior tract and trace program in place. (Figure 34-9)

Though the number of new cases was a fraction of what we saw in the southern US, Germany was concerned about excess demand on the healthcare system, international travel, the economy, and other matters.

Throughout August and September, Germany had 1800 to 2000 new cases per day. In the middle of August, tens of thousands of people protested in Berlin in defiance of self-mitigation rules.

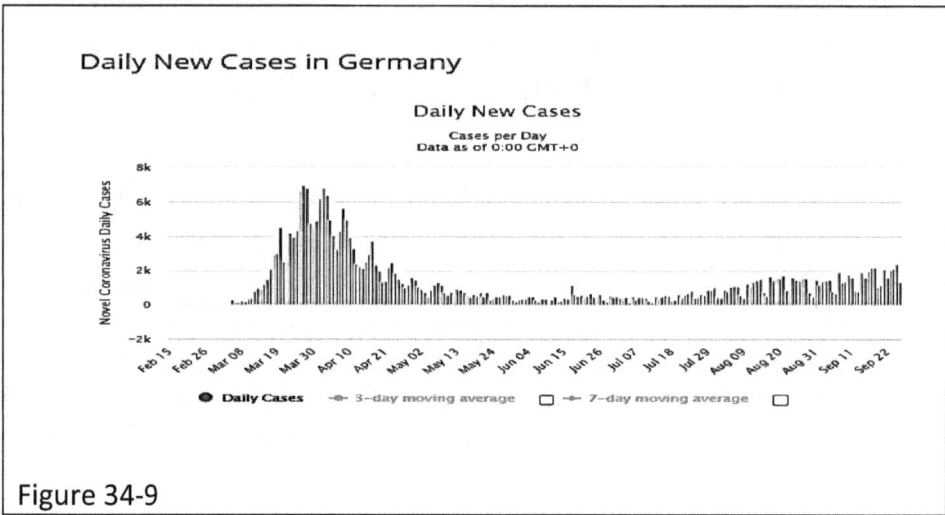

Figure 34-9

SPAIN

Spain also experienced a significant rise in the number of new cases. Their numbers were down in June and July. However, by the end of August, they had more than 10,000 new cases per day, in contrast to 2,000 new cases per day in Germany.

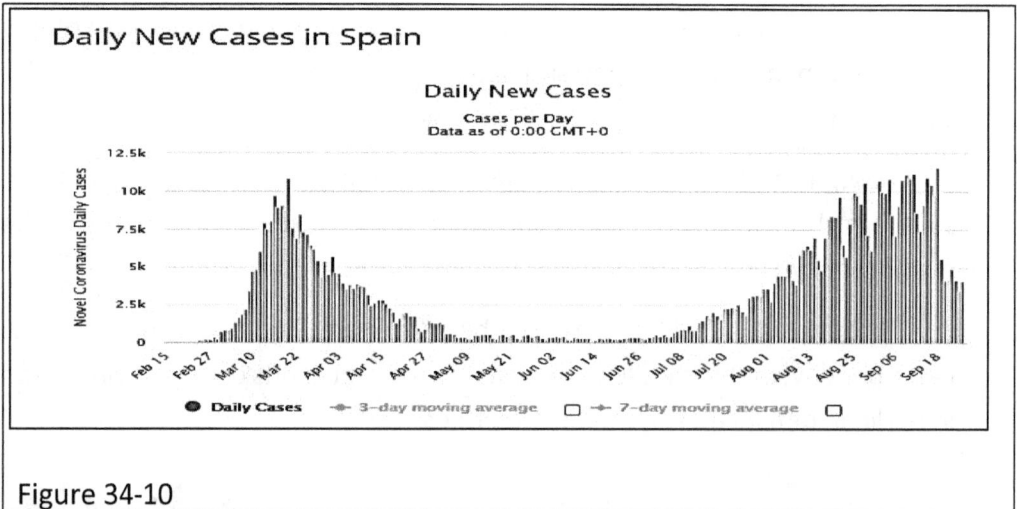

Figure 34-10

Spain's numbers declined in the second week of September. As observed in other countries, Spain's second surge impacted people aged 20-40 in contrast with the older population that was affected during the initial wave. (Figure 34-10)

THE UNITED KINGDOM AND FRANCE

In late September, a second wave raged across the UK and France. The number of daily new cases in the UK (6042) and France (14,412) were at their highest since the beginning of the pandemic in March 2020. In France, new cases were equivalent to the US' numbers, based on the population size. (Figure 34-11)

The UK was experiencing weekly gatherings of thousands of people defying the strict mitigation rules.

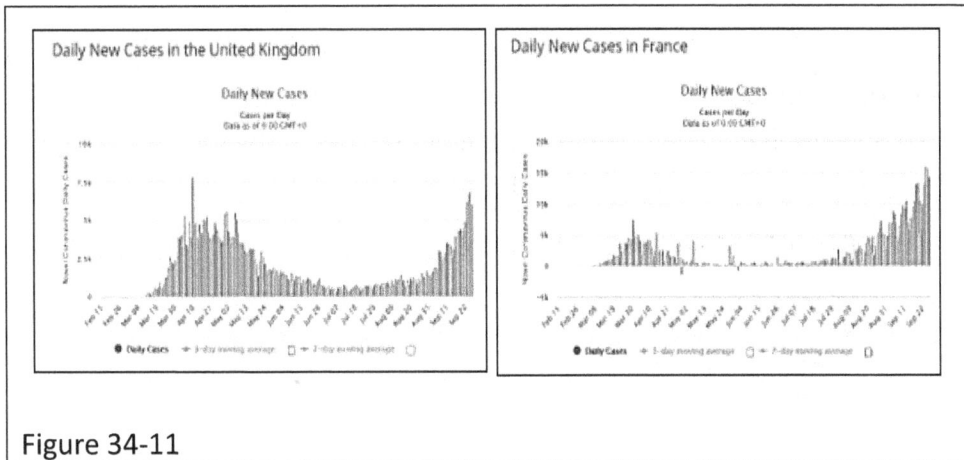

Figure 34-11

INDIA AND BRAZIL- STILL IN THE FIRST WAVE?

In August and September, India, which ranked number two, was still in the first wave while the rest of the world was experiencing a second wave.

Mumbai, a densely populated city (83,660 per square mile—second highest in the world) like New York City, had a raging spike in case numbers in September. Twenty-one percent of tested individuals were positive. (Figure 34-12)

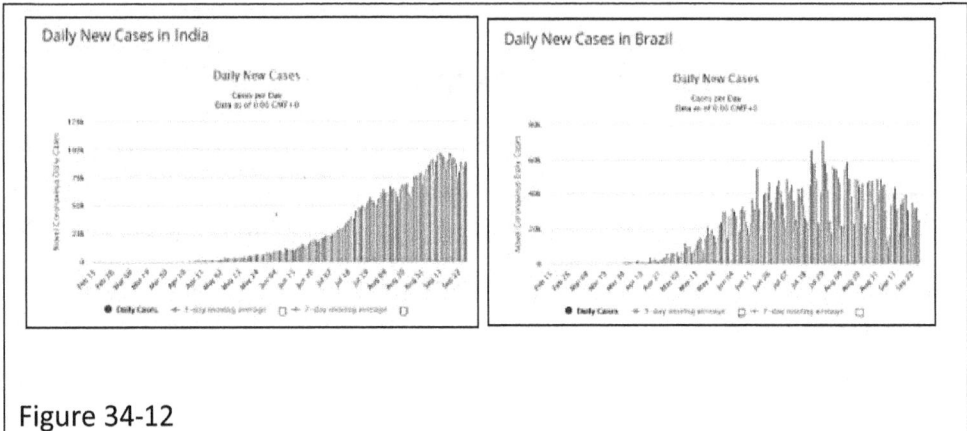

Figure 34-12

A comparable situation existed in Brazil, ranked third in the number of cases. However, Brazil had shown signs of declining numbers in recent weeks. Both countries had passed the lockdown period. Self-mitigation was the only hope until a vaccine became available.

LESSONS FROM THE SECOND WAVE?

- There is no set period within which a second wave or surge could occur
- The second wave could be more deadly than the first wave
- The second wave may overburden already-stretched healthcare resources
- Second waves test the people's patience, the policymakers' lack of knowledge and preparedness, and fear of a second lockdown
- As the second wave settles, there may be hope with more people exposed to the virus, further spread could be reduced
- Long-term self-mitigation is an important part of our lives after the first wave lift of lockdowns
- The deadly opportunistic virus spread around the world when people became more complacent

Between waves, it is a good time to:
- Stock face masks
- Stock hand sanitizers
- Keep medications for fever and aches

- Self-quarantine while waiting for test results

DID THE LOCKDOWNS DO ANY GOOD?

There was an expensive cost attached to each option.

Flattening the curve crushed the economy across the globe and rolled the misery down the timeline.

It was political suicide to avoid lockdowns when people were dying by the thousands, and the entire world was watching in despair.

A second lockdown was not a viable option when people were revolting and protesting the mitigation rules.

Some localized lockdowns and severe restrictions on gatherings were inevitable in many parts of the world, though very unpopular with people and politics.

If you add a volatile political climate, you can expect a rough ride until a vaccine comes in like a Knight in shining armor!

Ref:

https://thediplomat.com/2020/08/japan-braces-for-looming-second-wave-amid-dramatic-spike-in-covid-19-cases/

https://www.news.com.au/world/pacific/nz-cluster-sparks-fears-virus-arrived-on-international-flights/news-story/73d6432722dde6819309cf850a8a294b

https://www.thaipbsworld.com/can-thailand-escape-a-second-wave-of-covid-19/

https://www.washingtonpost.com/opinions/2020/08/10/japan-could-be-brink-second-wave-will-shinzo-abe-act/

https://www.dw.com/en/germany-corona-second-wave/a-54432548

https://www.nytimes.com/2020/08/04/world/australia/coronavirus-melbourne-lockdown.html

SECOND WAVE IN OTHER PARTS OF THE WORLD

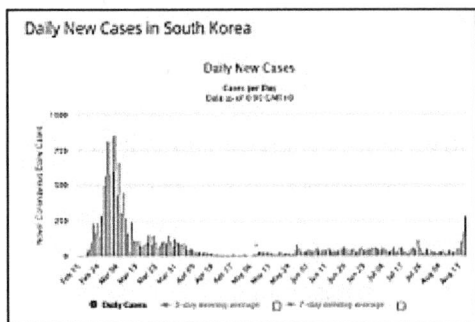

Daily New Cases in South Korea

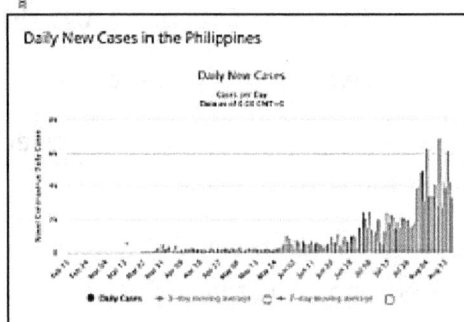

Daily New Cases in the Philippines

Daily New Cases in Italy

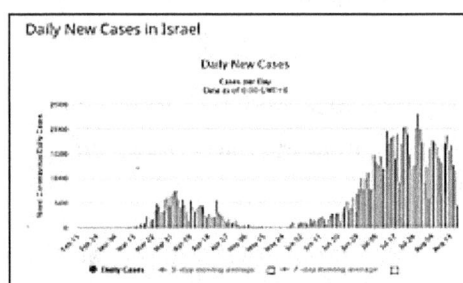

Daily New Cases in Israel

Daily New Cases in Sweden

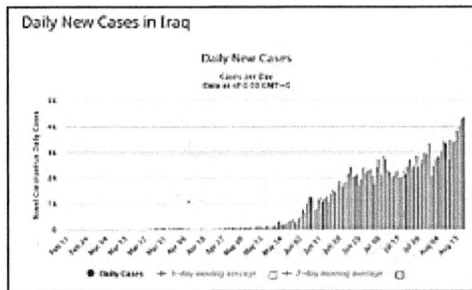

Daily New Cases in Iraq

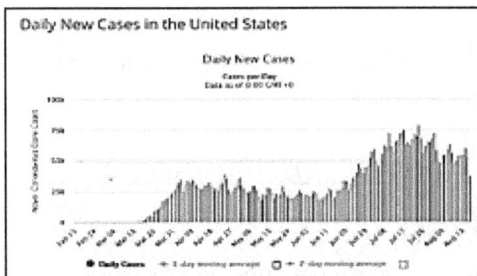

Daily New Cases in the United States

A Lesson from the Spanish Flu
Many thought the worst was over when the Spanish Flu pandemic of 1918 slowed with the arrival of warmer weather. But the chill of fall caused cases to skyrocket.

35 POST LOCKDOWN PARADIGM SHIFTS

AN UPTICK IN CORONAVIRUS CASES POST-LOCKDOWN

Soon after Texas lifted the lockdown on May 1, there was an increase in the new cases, which could represent a culmination of several factors:

- More people were getting tested as tests were readily available
- People we screened for COVID-19 before their surgical procedures and travels
- Most hospital inpatients got tested for COVID-19 to prevent the healthcare workers from being exposed unknowingly
- Some workplaces wanted their returning employees to get tested
- Holiday parties, beaches, and bars became breeding grounds for the rapid spread of the coronavirus
- Thousands of people challenging the curfew and joined the protests, defying the social mitigations

Younger people felt they were less likely to get a serious infection. However, during the second spike in Houston and other major cities in Texas, people aged 20 to 40 years accounted for 40% of hospital admissions. Also, younger people might spread the virus to the more vulnerable older members of the family.

UPTICK IN HOSPITAL ADMISSION AND ICU DEMAND

A dramatic increase in COVID-19 hospital admissions leading to an increase in demand for ICU beds and galloping consumption of healthcare resources was a major concern.

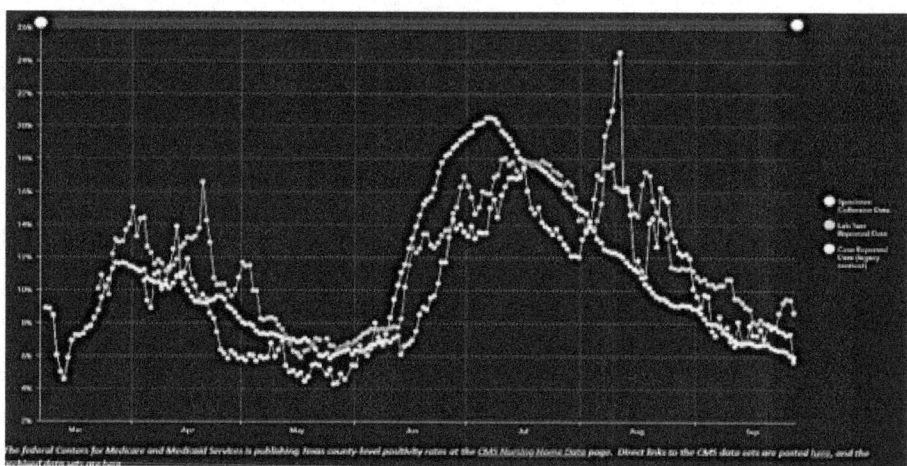

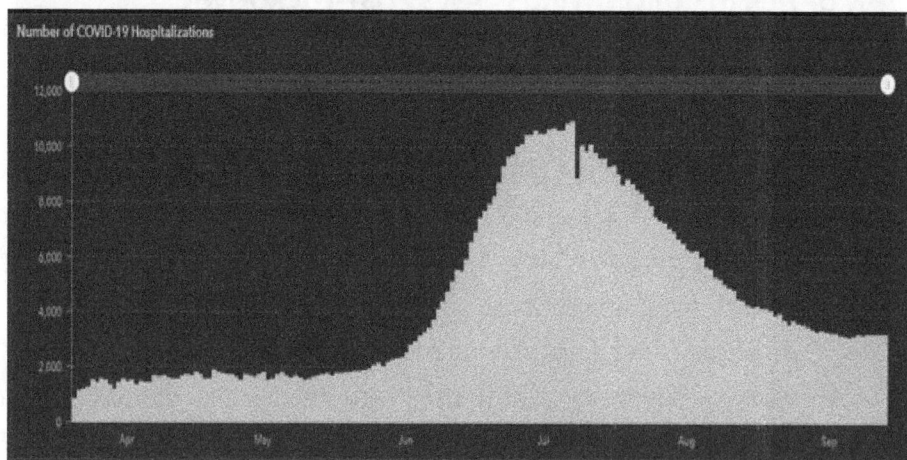

Figure 35-1

Since June 1, there was a steep increase in hospital admissions in Houston. Within a few days, hospital COVID-19 cases doubled, jumping to 10,000 by the end of July. At the end of September, numbers remained at 3000, which was 50% more than during April and May. (Figure 35-1)

In addition to a dramatic increase in hospital admissions, there was an alarming rise in the infection rate. The percentage of people testing positive in May was less than 4%. That number shot up to 21% at the beginning of July 2020. One out of every five people tested had the COVID-19 infection.

HOW DID THE UNITED STATES FARE DURING THE SECOND SPIKE?

The US was better prepared and equipped in June and July compared with February and March 2020.

- More and better testing capabilities
- PPE stockpiles were able to meet the increased demands
- There was a better understanding of the pathogenesis and course of COVID-19 infection
- Early interventions with Remdesivir and plasma transfusion showed better outcomes
- Hospital capacities were adequate to handle a sharp increase in ICU needs and healthcare resources, which were reaching a tipping point. However, there was no urgency to create temporary hospital beds

WHEN AND HOW DO WE INTERCEPT THAT COVID-19 POST-LOCKDOWN PARADIGM SHIFT

How to move forward from lockdown was the most challenging question for medical professionals, epidemiologists, and officials.

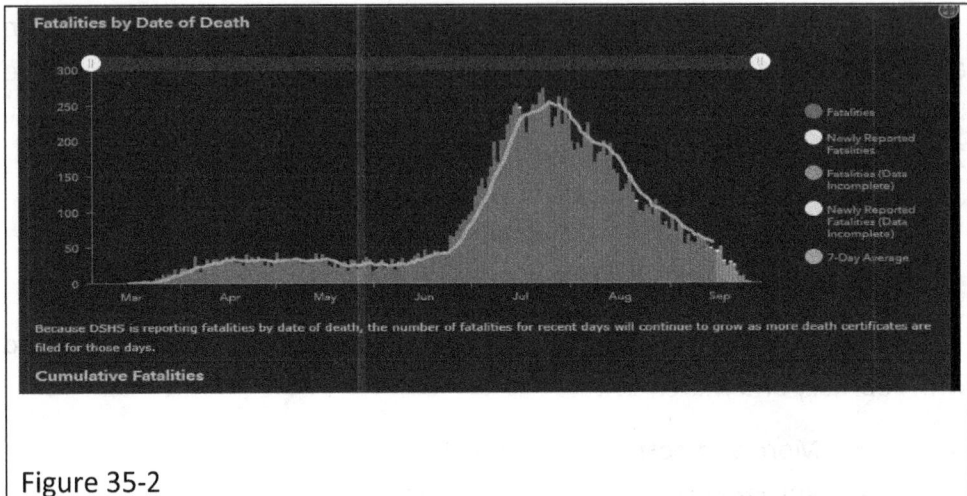

Figure 35-2

Facts, not media hype and sensational headlines, are critical.

The increase in COVID-19 cases could be due to more testing, an increasing percentage of people getting the infection or a combination of both. That was the case in Texas and other southern states. (Figure 35-2)

The most reliable number to follow is the number of people in the hospital and ICU units, which has a direct bearing on the demand placed on the healthcare resources and COVID-19 outcomes.

The death rates lag two to three weeks behind the active cases. Many variables could alter the numbers.

Interestingly, death rates progressively declined despite sustained hospital admissions, reflecting a better treatment outcome.

In Houston, the ICU utilization for COVID-19 patients was at 14.6%; the upper limit for ICU beds set by Gov. Abbott was at 15%. That signaled more demand for ICU beds for COVID-19 patients, which meant a reduced number of ICU beds for regular patients. The story was similar for regular inpatient beds.

WHAT SHOULD BE DONE?

The only way to slow the viral spread was to identify active cases, isolate them, track their contacts, and educate the public.

That proposition sounded great in theory but was very time-consuming, involving the services of thousands of caseworkers, huge financial commitment, and an understanding and cooperative community.

From a practical point, those variables were beyond the abilities of major cities. The most viable option came back to self-mitigation at individual levels.

People knew what they needed to do to stay safe. Getting them to follow these guidelines for extended periods was another story. Yet, that concept was vital for schools, universities, and businesses when they detected active cases at their locations.

OUTCOME OF LOCKDOWN

Did lockdown reduce the spread of infection to other people? People questioned the value of lockdown when New York City reported that over 60% of admissions to hospitals involved people were in lockdown.

Living in a high-rise apartment complex was very much like living in a closed environment like a nursing home. They were exposed to visitors, the common hallways, air condition, the elevators, etc. So, what was the alternative?

People defying the lockdowns and taking to the streets in mass protests may have contributed to the increased case number. Mixing in public in large numbers presents a massive risk—Sweden, which had extremely limited mitigation, had one of the highest mortality rates per million population.

The rules and regulations in different states reflected different ideologies based on their political alliances. That created friction between the federal administration and the states.

TELECONFERENCING

Teleconferencing became a new normal, replacing face-to-face company meetings. Some businesses adopted work-from-home permanently.

VIRTUAL WORLD

We saw graduations going online and people being receiving their graduation confirmations over internet calls. The year they would remember forever!

TRAVEL TROUBLES

The pandemic brought air travel to a virtual standstill. In October, as a possible second wave was sweeping through the world, it looked like it would be several months or a year before air travel would again populate the skies.

RESTAURANTS

Many restaurants had to shut their doors permanently because they could not survive with 25%-50% occupancy and no catering orders. Virtual meetings and functions do not generate any catering.

RETAIL BUSINESSES

We saw retail stores dying one by one (e.g., JCPenney). Even if the malls opened, people did not have the money spend. The malls, which had no air-conditioning for two months, also had to address molds and dust problems.

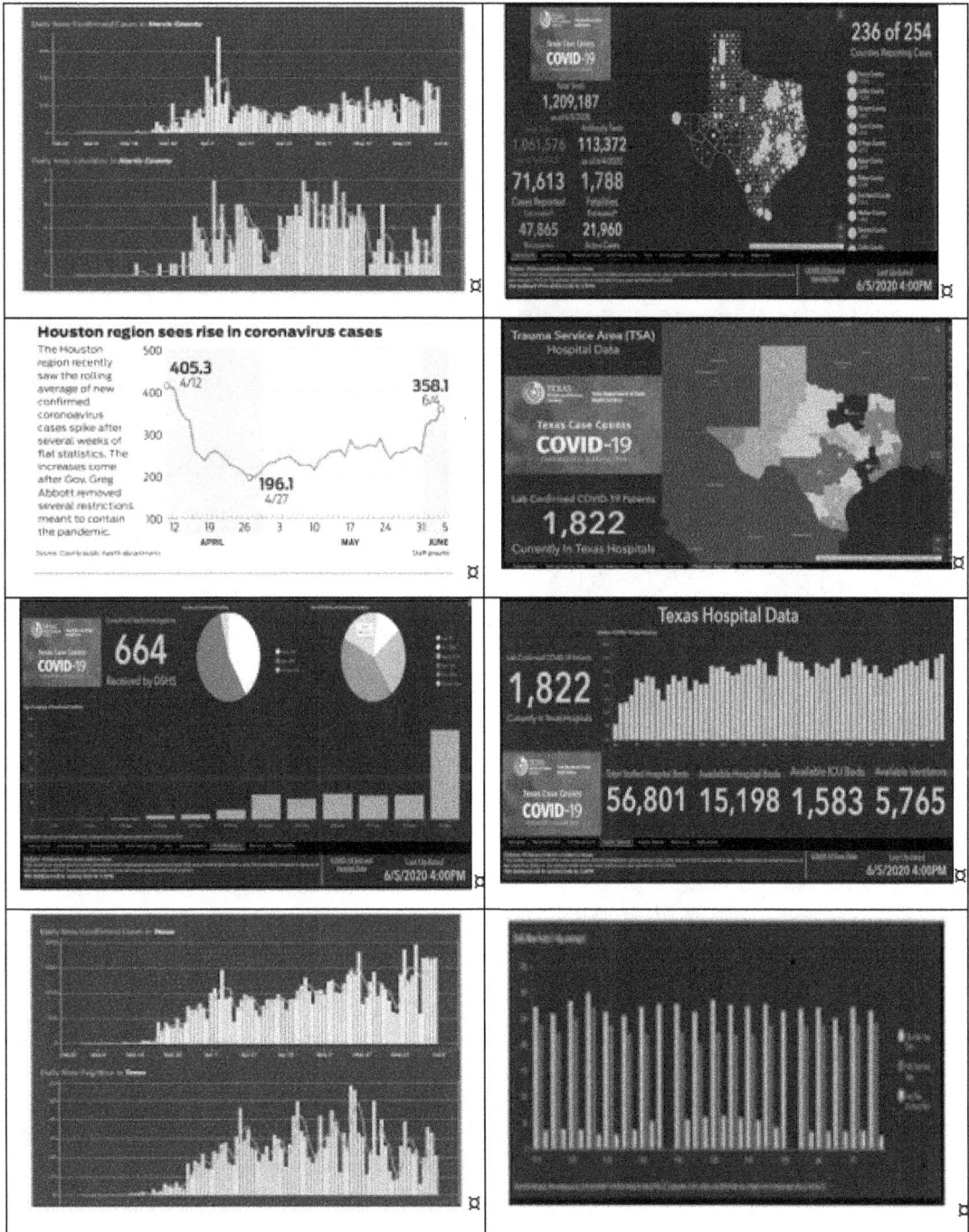

Figure 35-3

Nik Nikam
June 4 at 10:03 PM · general information/re...

SWEDEN AT CROSS ROADS WITH CORONAVIRUS PANDEMIC

Nik Nikam, MD, MHA. HOUSTON, TEXAS.

Sweden is a Scandinavian nation, with Stockholm as its capital. It has a population of 10.23 million people (2019). It is surrounded by Norway on the west and Finland on the east, Poland and Germany on the south.

Sweden is one of the few countries that allegedly adopted a very relaxed quarantine measures, compared to the strict lockdown seen in other EU countries. Its relaxed policies maybe be bl...
See More

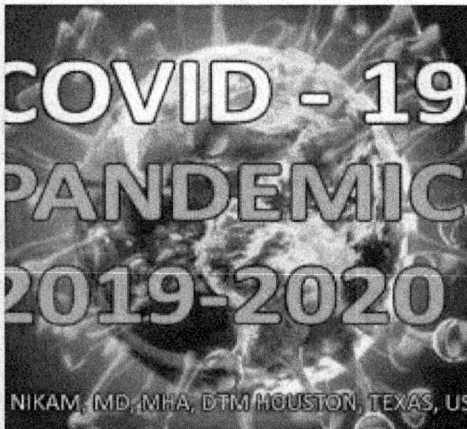

15,959	33,204	172
N/A	N/A	116
20,171	24,697	92
N/A	N/A	308
20,079	11,735	56

Daily New Cases in Switzerland

aily New Cases in Sweden

+8

👍❤️😮 62   116 Comments

👍 Like   💬 Comment

36 COVID-19 PANDEMIC TO CASEDEMIC IN OCTOBER?

WORLD

On Oct 17, 2020, over 39 million cases of COVID-19 were reported globally, with more than 1.1 million deaths. That represented a little over 2% of the 7.38 billion people on this planet. It also meant that 97% of the population was potentially vulnerable to COVID-19 infection. If we assumed about 40% were patients were asymptomatic, the number of cases (38 M X 1.4) accounted for would still be less than 56 million, which left 92.4% of the population susceptible to the coronavirus infection. (Figure 36-1)

WORLD / COUNTRIES / BELGIUM
Last updated October 18, 2020, 13:19 GMT

||| Belgium

Coronavirus Cases:
213,115

Deaths:
10,392

Recovered:
21,074

Figure 36-1

WAS COVID-19 MAKING A THIRD PEAK IN THE UNITED STATES?

Just when we thought the second peak was finally showing signs of relief, there was a resurgence of new cases in the US for the third time!

The number of new cases in the US was approaching 75,000 per day, and it just looked like the upstroke of the third Farr curve! Even though the US recorded double the number of new cases during July and October, the death rates have not doubled. The death rate in July and October was less than 50% of what it was in April 2020. (Figure 36-2)

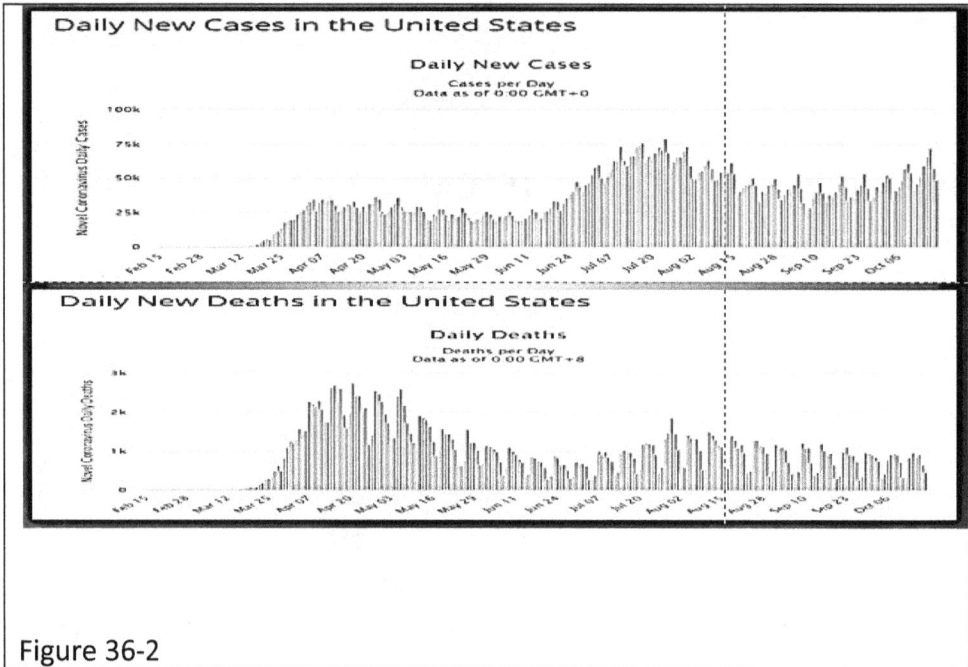

Figure 36-2

EUROPE

In the EU block, France is reporting 30,000 new cases per day, which was three times the number it reported in March and April 2020. With a significant increase in the number of new cases, the mortality rates were on the decline.

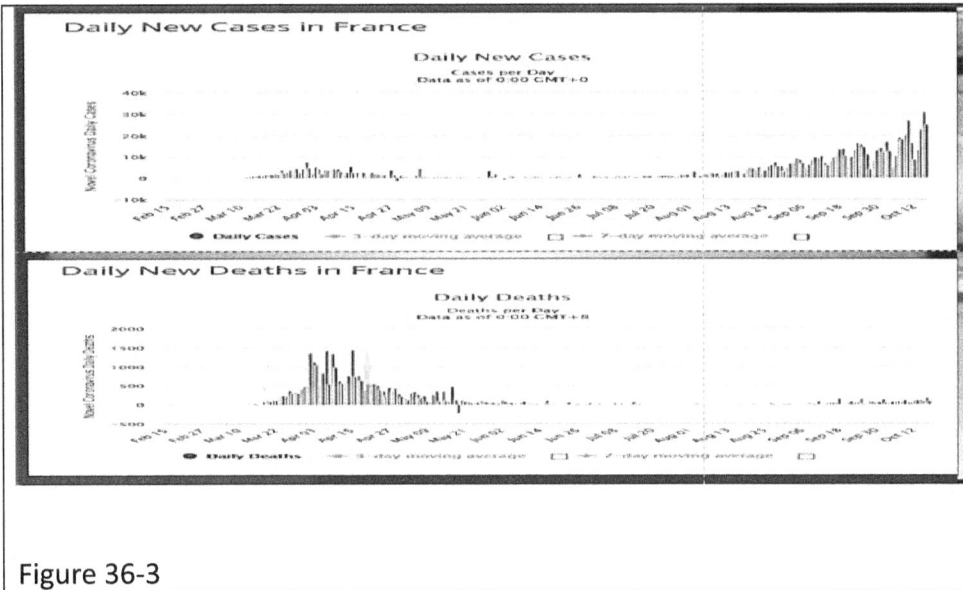

Figure 36-3

GERMANY

Germany, with the best COVID-19 pandemic strategy, reported 7,500 new daily cases, surpassing the numbers seen in spring. But the death counts were nowhere the numbers noted during the first wave. (Figure 36-4)

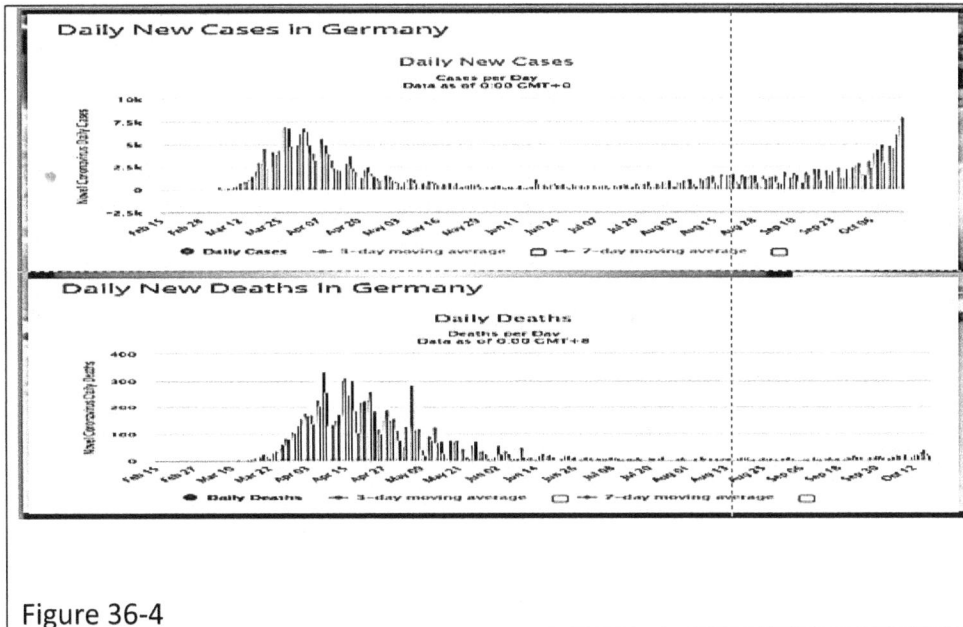

Figure 36-4

JAPAN

Japan's second wave, which recorded 1500 new cases per day at the peak of its second (Max 500 cases during the first peak in April), is reducing now with 500 new cases per day. (Figure 36-5)

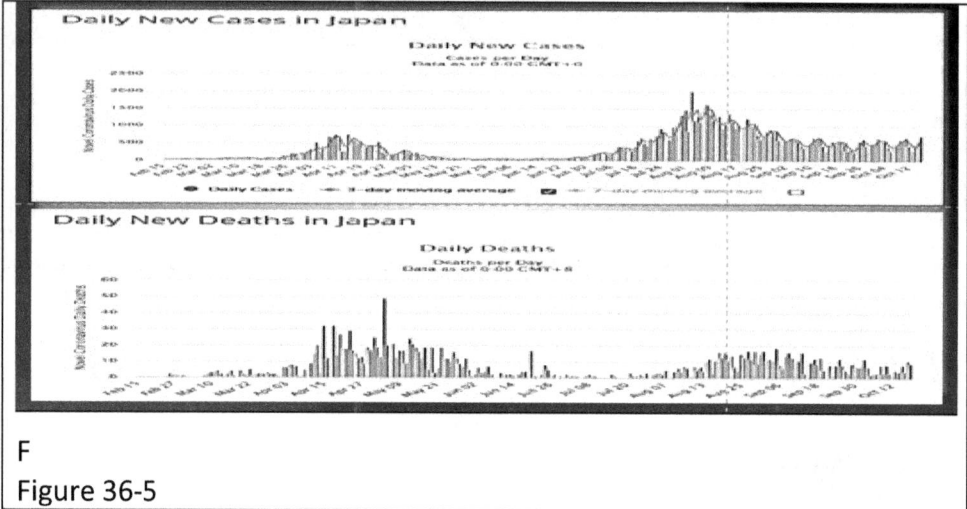

F

Figure 36-5

BELGIUM

The figure from Belgium showed a similar response. The death rates were a fraction of that during the April peak, despite the new cases triple in the meantime. (Figure 36-6)

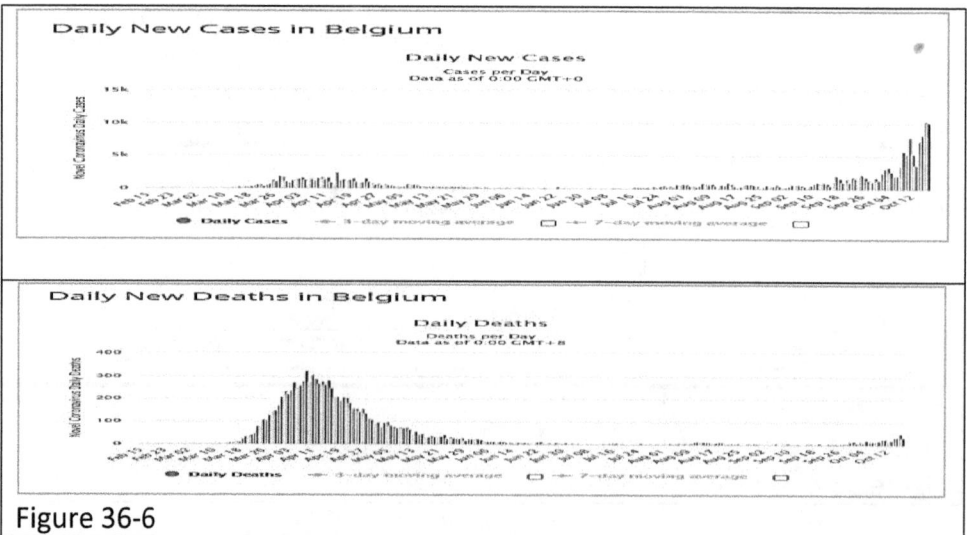

Figure 36-6

FRANCE

France mirrored the response seen in Japan, Germany, and Belgium with dubbing and tripling of number with declining death rates. (Figure 37-7)

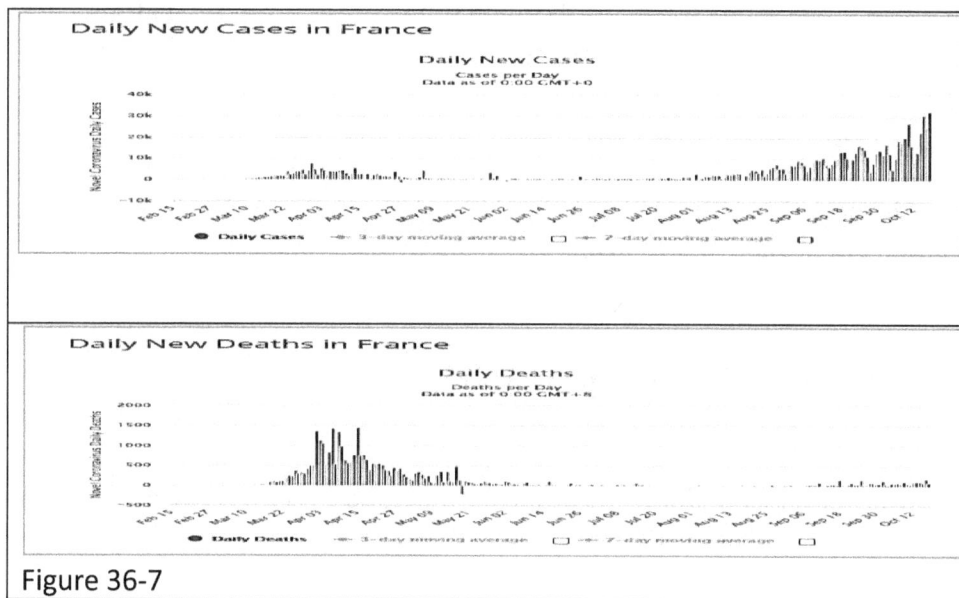

Figure 36-7

"CASEDEMICS" WITH DRAMATIC DECLINE IN DEATH RATES

The figures from Germany, France, and Belgium underscored the changing landscape of the COVID-19 pandemic. Though alarming to the naked eye, a deeper scientific analysis would unravel the mystery and the dichotomy between declining death rates despite an astronomical rise in new cases in the same countries.

The COVID-19 pandemic is just picking up pace, penetrating deeper into the communities across the world, sparing no country, no matter how good their testing and contact tracing systems were!

As we learned, the virus could be airborne, and the chances of virus exposure were more than we had initially thought.

Younger people were also getting the infection, though their mortality rate was extremely low (less than 0.1% for those under 29 years)

As the pandemic lingers in various stages in different parts of the world, the after-effects of lockdown and economic toll, especially on the lower-income groups are very palpable.

As people swung from one extreme to the other; from total lockdown to total disregard for the self-mitigation rules, the emotional disruption, the income loss, and the lack of social interaction were bound to have long-lasting repercussions.

The new case rates have gone up in many countries, while the death rates continue to decline in proportion to the new case numbers.

WHAT COULD EXPLAIN THE CASEDEMIC DICHOTOMY?

More tests were done in a population that had very mild symptoms. The increase in the percentage of cases testing positive might not be alarming if the death rates remained low.

The PCR tests detected remnants of viral RNA, which may not equate to active infection, and thus spuriously increasing the numbers without real consequences or outcomes.

There was increased political pressure to test more and more people, in some parts of the world

37 A ROADMAP FOR FUTURE PANDEMICS

WHAT TO FOCUS ON?

Life isn't the same. So, what's left? You can't travel, you can't go to a restaurant, you can't party, and you don't want any guests at your house. Does this mean you can't have fun?

"The only thing that's left is your life, and this is the most important asset you have! You need to protect life at all costs until the pandemic fades away or you get a vaccine! Quit complaining!" they told me!

Since pandemics run two to three waves over 18 to 48 months, it may make life more complicated and challenging during fall and winter 'flu seasons. It may drain the healthcare resources.

The best thing you can do is to protect yourself until a definite treatment or vaccine is available. This means that whatever self-mitigation means you followed during lockdowns should be continued as you return to work. Now is a good time to pick up a hobby that brings joy while allowing for safe practices and try to get absorbed in that for the next 6 to 12 months.

We learned many important lessons from the COVID-19 pandemic. Here is a partial list of life-saving tips for future pandemics:

✓ Early detection
✓ Early quarantine
✓ Self-mitigation from day one. If there is a virus in your town, you could be next. Follow self-mitigation rules and wear a mask. Don't wait for the experts or politicians to tell you!
✓ Stop denying or finger-pointing
✓ PPE
✓ Hospital beds
✓ Economic support
✓ Financial support
✓ Exchange of information across the globe
✓ Standardization of recording and reporting data.
✓ Early travel restriction
✓ Virus detection kits before the invisible enemy comes to town

THE MOST IMPORTANT NUMBERS THAT MATTER TO MEDICAL PROFESSIONALS AND HEALTHCARE LEADERS?

1. Total number of daily new cases

2. Total number of new admissions to hospitals

3. Total number of patients on ventilators

4. Total number of new patients needing ventilators

5. Total number of recent daily deaths

6. Total number of patients in the ICU and ICU bed capacity

7. Case doubling rate

Bases on these realistic facts, you can access the current situation, healthcare capacity, and future needs in the coming days.

CONCLUSION

As I carve out the last words in this book, I can't help but think of the wild, frightening times we are living in, and the dizzying roller coaster ride the force of nature has inflicted on us.

SARS-CoV-2. A virus no bigger than one-micron that has no brain, no feelings, no army, and no kingdom. A virus whose only purpose is to invade human beings and destroy their lungs as it replicates itself to spread to many more people.

As of Oct 23, 2020, over 40 million people had contracted the virus, and over 11.1 million deaths had been recorded worldwide. The worst may be yet to come, according to science and health experts.

The Spanish 'Flu had three waves spanning 18-24 months. The second wave was five times as deadly as the first. In the end, it took over 50 million lives.

COVID-19 may not reach the same magnitude due to a better understanding of the virus and dramatic mitigation measures implemented during the first wave. Still, SARS-CoV-2 caused widespread human misery, isolation, economic calamity, and an uncertain future that may have sociopsychological effects that could linger for decades. The most affected are likely to be society's lower economic echelon if a vaccine is not available soon.

Health experts have predicted the death toll in the US could exceed 400,000, in the coming months, despite all our efforts. This death toll may march on like the titanic, independent of human efforts and influences until we have a drug or a vaccine that can choke this virus.

I am optimistic that we will have a vaccine that will change the course of the COVID-19 pandemic for the better in the coming months. The entire world is waiting!

People will talk about the COVID-19 pandemic for generations to come. COVID-19 will join the ranks of the bubonic plague and the Spanish 'Flu, at least in terms of human anguish and its economic impact on the world.

Stay Tuned for The Rest of the Story!

The birth place of COVID-19 PANDEMIC BOOK & VIDEOS (2020)

Someone had to do the job

Once upon a time, long before COVID-19 visited our town

www.ingramcontent.com/pod-product-compliance
Lightning Source LLC
Chambersburg PA
CBHW080046280326
41934CB00014B/3236